ADVANCED INORGANIC CHEMISTRY

ADVANCED INORGANIC CHEMISTRY

Arun Rastogi

ANMOL PUBLICATIONS PVT. LTD.
NEW DELHI-110 002 (INDIA)

ANMOL PUBLICATIONS PVT. LTD.
Regd. Office: 4360/4, Ansari Road, Daryaganj,
New Delhi-110002 (India)
Tel.: 23278000, 23261597, 23286875, 23255577
Fax: 91-11-23280289
Email: anmolpub@gmail.com
Visit us at: www.anmolpublications.com

Branch Office: No. 1015, Ist Main Road, BSK IIIrd Stage
IIIrd Phase, IIIrd Block, Bangalore-560 085 (India)
Tel.: 080-41723429 • Fax: 080-26723604
Email: anmolpublicationsbangalore@gmail.com

Advanced Inorganic Chemistry

First Edition, 2010

ISBN 978-81-261-4372-6

PRINTED IN INDIA

Printed at Mehra Offset Press, Delhi.

Contents

Preface

The textbook "Advance Inorganic Chemistry, is the study of all the elements and their compounds with the exception of carbon and its compounds, which fall under the category of organic chemistry. Inorganic chemistry is significant in all modern areas of life and science. It represents one of the fastest growing and exciting areas of scientific research. The impetus for this research is the practical essential role inorganic chemistry plays in biology, pharmaceuticals, and advanced materials.

This book explores the latest advances in the field of inorganic chemistry. It is intended for the undergraduate or introductory graduate inorganic chemistry course of sequence of courses, such as the one-semester junior-senior level advanced inorganic chemistry course."

Author

Chapter 1

Water and its Hardness

HARD WATER

We're not talking about frozen water (ice). Calcium and magnesium ions (these are both positive ions) in water join with the soap to form as insoluble curd and reduce the effectiveness of the soap as a washing agent.

The real problem comes when you try to wash the curd out of the clothes because as an insoluble material it sticks to the fibres of the clothes, does not re-dissolve and is left behind after rinsing.

So you may have added dirt to the clothes (in the form of these insoluble calcium and magnesium compounds) rather than cleaning them.

Clothes may feel damp even though they are dry, and do not have a 'crispness' after washing. Over time, this soap curd will give a dull yellow colour to white materials.

In the bath or shower, soap is difficult to get a lather with hardwater and instead of feeling slippery it seems difficult to wash. Your hair, in particular, starts to matt together and it's difficult to run your fingers through your wet hair.

When your hair dries it feels sticky and clings together because the calcium and magnesium compounds have not been removed from the hair fibres but are acting like glues. Your skin may be clammy because the calcium/magnesium precipitate absorbs moisture, and may also leave the skin drier than usual.

In the bath, there are tell-tale signs of the high water mark and the hand basin gets a greasy film over the surface of the porcelain. It requires cleaning with some of the special products because this scum is insoluble in water. Shower screens and tiles get a matt appearance (dull) and appear to have a coating that will increase in thickness over time, unless removed.

It is usual to use a proprietary product to remove this 'grunge'. These products are usualiy acidic to dissolve the calcium and magnesium precipitate.

Inside pipe, hard water leaves an insoluble deposit of calcium and magnesium carbonates and over time may completely clog up the pipe.

In hot water services, in the electric kettle or coffee making machines, the calcium deposits on the heating element and can build up to a stage that the water is no longer in contact with the element, the element overheats and explodes.

MEASURING HARD WATER

There are some quick and simple test kits used to test for hardness of swimming pool water. In the laboratory, the water is analysed for calcium and magnesium and a calculation is made. Hardness can be reported for calcium hardness, magnesium hardness or total hardness (the sum of the calcium and magnesium hardness). Total hardness is reported in milligrams per litre of calcium carbonate equivalent (mg/L $CaCO_3$).

Some imported water softeners refer to hardness in obsolete German units of "degrees of hardness" (^{o}dH). To convert ^{o}dH to total hardness in mg/L $CaCO_3$, divide by 0.14.

CLASSIFICATION OF HARDNESS

The Australian Drinking Water Guidelines state that there are no health guideline values for total hardness, but an aesthetic value is suggested at 200 mg/L $CaCO_3$ (which is reasonably hard water).

Very hard water is likely to cause scale (insoluble calcium and magnesium compounds) to form on the inside of pipes and boilers.

- Soft < 60 mg/L $CaCO_3$

- Good quality 60-200 mg/L $CaCO_3$
- Increasing scaling problems 200-500 mg/L $CaCO_3$
- Severe scaling >500 mg/L $CaCO_3$

Other classification systems are intermediate to the Australian Drinking Water Guidelines and refer to various scales of total hardness as:

Soft	<17 mg/L $CaCO_3$
Slightly hard	17–60 mg/L $CaCO_3$
Moderately hard	60–120 mg/L $CaCO_3$
Hard	120–180 mg/L $CaCO_3$
Very hard	>180 mg/L $CaCO_3$

DRINKING WATER SUPPLIES

It is usual in many metropolitan and urban areas for the water supply authorities to reduce the hardness of the water from their reservoirs so that its aesthetic quality is improved from the raw water resource.

Reducing hardness is simply a matter of removing some of the calcium and magnesium ions, not all of them, just enough to make the water softer.

Since calcium and magnesium rapidly form precipitates with carbonates, adding soda ash (sodium carbonate) to water forms calcium carbonate and magnesium carbonate that then settles out as a precipitate.

Soda ash is also called 'washing soda' and its use for normal laundry purposes is to reduce the calcium and magnesium effects on soap. In detergents, phosphorus salts are used to isolate the calcium and magnesium in a different way (discussed later).

I measured 43 town water supplies across coastal and inland NSW as part of a research project. The hardness in the water supplies was extremely variable but noticeable different between the coastal fringe and the inland (west of the Great Dividing Range).

The number of towns (frequency distribution) for a range of total hardness. It is clear that inland water supplies are harder than coastal supplies. As a comparison, rainwater has a hardness of less than 5 mg/L.

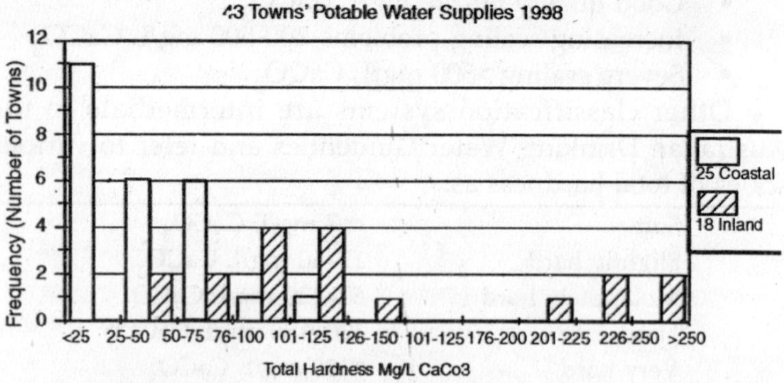

METROPOLITAN WATER SUPPLIES

A search of the internet revealed the following levels of total hardness in metropolitan water supplies

Table. Total Hardness in Metropolitan Water Supplies

Metropolitan Supply	Total Hardness (mg/L $CaCO_3$)
Melbourne - City West Water	15-29 mg/L
Melbourne	10-26 mg/L
Sydney's water	50-60 mg/L
Brisbane	119-129 mg/L from Mt Crosby
Brisbane	69-77 mg/L from North Pine
Adelaide	190-323 mg/L from Anstey Hill
Adelaide	262-442 mg/L from Hope Valley

EFFECTS OF HARDNESS ON SOAPS AND DETERGENTS

The problem that hardness causes to our use of water for washing (persons and clothes) is that the calcium and magnesium salts react with the surfactant (the surface active agent) and reduce its effectiveness. We can overcome that problem by using more washing soap to bind the calcium and magnesium.

Additional amounts of washing soda (sodium carbonate) or borax can be used to precipitate the calcium and magnesium.

The problem with soaps in hard water is that the compounds formed with the soap result in a greasy curd that is insoluble in water and difficult to wash out. It doesn't matter how much extra soap we use, the curd remains insoluble.

So if we have to learn to live with hard water, how do we overcome the effects of the hardness. The harder the water, the more detergent you use.

Phosphates have been added to water to deal with hardness and these salts are very good at combining with calcium and magnesium and keeping them in suspension so that they can be effectively washed out of the clothes.

Whether substitutes for phosphorus, such as zeolites work in hard water may need to be tried. The third way of dealing with hard water is to pass the water through a softener.

WATER SOFTENERS

These devices are ion exchange systems - that is, they exchange sodium ions for calcium and magnesium ions. Water passes through a softener, usually a large tank something like a LPG gas bottle.

On its way through the media inside the softener, the calcium and magnesium ions are captured by the media and a replacement occurs.

Sodium ions on the media are displaced by the calcium and magnesium. Instead of the water having say 100 mg/L (parts per million) of calcium and magnesium combined, the softened water now has very low levels of calcium and magnesium, but levels of sodium displaced are fractionally higher than the combined calcium and magnesium displaced.

At regular intervals, the ion exchanger has to be replenished, that is back flushed with a brine solution (concentrated sodium chloride solution) to replace the calcium and magnesium so the process can start again. The backwash should be discharged to sewer and not onto the ground or into a septic tank (or other on-site system).

The period between backwash events will depend upon the hardness of the water and the volume of water treated. Check with your irrigation specialist for the correct sizing of a water softener.

Reverse Osmosis (RO) is the removal of contaminants from water by forcing the water through a semi-permeable membrane.

These devices produce very clean water, but are expensive to operate and waste about 90% of the water passing through them. RO water is'too pure' to drink.

CONVENTIONAL WATER SOFTENING

Most conventional water-softening devices depend on a process known as ion-exchange in which"hardness" ions trade places with sodium and chloride ions that are loosely bound to an ion-exchange resin or a zeolite (many zeolite minerals occur in nature, but specialized ones are often made artificially.)

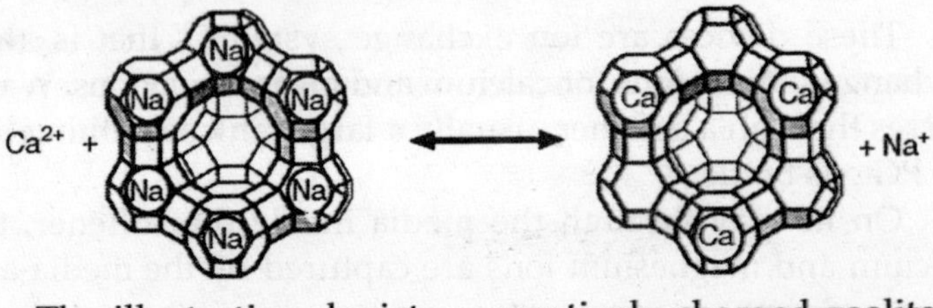

The illustration depicts a negatively-charged zeolite to which [positive] sodium ions are attached. Calcium or magnesium ions in the water displace sodium ions, which are released into the water.

In a similar way, positively-charged zeolites bind negatively-charged chloride ions (Cl^-), which get displaced by bicarbonate ions in the water. As the zeolites become converted to their Ca^{2+} and HCO_3^- forms they gradually lose their effectiveness and must be regenerated.

This is accomplished by passing a concentrated brine

solution though them, causing the above reaction to be reversed.

Herein lies one of the drawbacks of this process: most of the salt employed in the regeneration process gets flushed out of the system and and is usually released into the soil or drainage system- something that can have damaging consequences to the environment, especially in arid regions.

For this reason, many jurisdications prohibit such release, and require users to dispose of the spent brine at an approved site or to use a commercial service company.

TYPES OF WATER HARDNESS

TEMPORARY HARDNESS

This refers to hardness whose effects can be removed by boiling the water in an open container. Such waters have usually percolated though limestone formations and contain bicarbonate HCO_3^- along with small amounts of carbonate CO_3^{2-} as the principal negative ions.

Boiling the water promotes the reaction:

$$2\,HCO_3^- \rightarrow CO_3^{2-} + CO_2$$

By driving off the carbon dioxide gas. The CO_3^{2-} reacts with Ca^{2+} or Mg^{2+} ions, to form insoluble calcium and magnesium carbonates which precipitate out. By tying up the metal ions in this way, the amounts available to form soap scum are greatly reduced.

PERMANENT HARDNESS

Waters than contain other anions such as chloride or sulfate cannot be remediated by boiling, and are said to be "permanently" hard.

TREATMENT OF HARD WATER

WATER TREATMENT METHODS

There are many commercial ways of treating hard water including water filters, water softeners, electromagnetic water conditioners and reverse osmosis. In this section we will look at the pro's and con's of several of the available methods to treat hard water.

Method	Chemical Water Softeners	Water Filters	Mechanical Water Softeners	Magnetic Water Conditioners
Description	Chemicals added in laundering process	Drinking water from the tap is filtered with a portable unit	Permenantly installed in plumbing system to replace calcium and magnesium with sodium	Permenantly installed in plumbing system to alter calcium ions so they cannot cause limescale
Advantages	Removes Calcium and Magnesium and lengthens lifespan of clothes	Output water is fit for drinking Chlorine removed Water Softened	Well Established Techology Prevents Limescale Increases heating efficiency Lengthens lifespan of clothes	Output water is fit for drinking Calcium retained, good for diet Low purchase and installation cost (easy DIY install) Low running cost
Disadvantages	Output water is not fit for drinking	Does not prevent limescale deposits in pipes and heating system	Output water may not be suitable for drinking Expensive	They don't work for everybody Some skepticism over their effectiveness
Approximate Cost		< £0.03 ($0.05) / litre	£1000 ($1400) Unit + Installation £100 ($140) / Year - Salt + Electricity	£100 ($150) Unit + Installation £5 ($7) / Year - electricity
Recommended Manufacturers		Brita Worldwide	Worldwide Water Systems	Water Imp

PACKAGED WATER SOFTENERS

The+ontent of softened water, some individuals may be advised by their physician, not to install water softeners, to soften only hot water or to bypass the water softener with a cold water line to provide unsoftened water for drinking and cooking; usually to a separate faucet at the kitchen sink.

Mechanically softened water is not recommended for watering plants, lawns, and gardens due to its sodium content.

WATER FILTERS

It may be useful to invest in a water filter for your home to make sure that your drinking water is clean and free from contaminants. There are different forms of water filters available today, all useful as a water softening treatment:

- *Granulated, activated carbon water filters*: These are portable filters that are found in water jugs. They are the most simplest type of filter and removes contaminants such as chlorine, large particles and parasites. They are available at most kitchenware stores and are reasonably cheap if you have a low budget. However, they have a short filter life and don't filter out many chemicals and bacteria.

- *Carbon block, activated carbon water filters*: These types of filters are generally built-in to the water supply in your home. These filters are more expensive than granulated types, but they do give better filtration and do not need to be replaced as often.
- *Ceramic carbon water filters*: This type of filter is comprised of a ceramic part and a carbon part, this gives it an ability to filter out a broad range of contaminants from the water supply. Some are infused with silver, as this is a good antibiotic and reduces the number of micro-organisms present. These filters are good value for money. They are built-in to the home water supply and need to be replaced about every 12 months in the average household.
- *Reverse osmosis water filters*: These filters are often considered the best form of water filtration available today. The water is passed through a semi-permeable membrane, which filters out almost every contaminant. They are more expensive than other types of filters but have reasonably low running costs, effective filtration and can be used to filter water to the whole house.
- *Combination Systems*: This filter has become available in recent years and are seen as an alternative to reverse osmosis systems. They have a number of different filters, and each filtration process removes different contaminants.
- *Alkaline Water Machines*: This system not only filters your drinking water but it makes it more alkaline. It is a good system for those with environmental

illnesses where there is too much acid in the body. It has effective filtration but it can be expensive and it wastes a lot of water during the filtration process.

MAGNETIC WATER CONDITIONERS

Electromagnetic water conditioners are a relatively new invention. The idea is that by passing water through a magnetic field, the calcium and magnesium ion's are altered in such a way that they loose their ability to cause scale.

This has a number of benefits; although the water is not technically soft, it has the useful properties of soft water, that is, it won't cause limescale in your pipes thus increasing heating efficiency and lengthening the lifespan of any clothes washed in the conditioned water.

Calcium is an important dietary element, so the fact that conditioned water still retains its calcium content is an added benefit. While some people are skeptical this method actually works, we have found one manufacturer who has commissioned a scientific analysis by the University of Bath that concludes that their device does indeed stop the build up of limescale.

HARD WATER PROBLEMS

Water described as "hard" means it is high in dissolved minerals, specifically calcium and magnesium. Hard water is not a health risk, but a nuisance because of its tendancy to cause mineral buildup in water pipe and heating systems, and its poor soap and/or detergent performance when compared with soft water.

Water is a good solvent and picks up impurities easily. When it combines with carbon dioxide in the air to form very weak carbonic acid, an even better solvent results.

As water moves through soil and rock, it dissolves very small amounts of minerals and holds them in solution. Calcium and magnesium dissolved in water are the two most common minerals that make water "hard." The degree of hardness becomes greater as the calcium and magnesium content increases.

Process

There are a number of tips you can follow to reduce the effects of hard water in your home, without having to make any major changes:

Choose a correct laundry detergent – Some laundry detergents do not produce as many suds in hard water, these are likely to be soap-based products and do not work as well in hard-water as detergent based products.

Nowadays, there are washing powders and liquids available for a wide range of water hardness. Make sure you choose the correct detergent for your area; you may also need to use slightly more detergent than the manufacturers recommended amount to compensate for the hard water. In many cases the manufacturer will give specific instructions on how to use the product in hard water areas, look out for these labels on your product.

Reduce the temperature of your boiler – As the water temperature increases, the more mineral deposits will appear in your dishwasher, water tank and pipes.

By reducing the heat of your boiler to about 55°C, you will have enough hot water for your shower and you will reduce the amount of mineral build-up in your pipes and tanks. Use rinse agents to remove mineral deposits – There are many rinse agents available to remove mineral deposits from crockery and dishwasher.

Alternatively, you can use white vinegar by using the dishwasher dispenser or placing a cup of vinegar on the dishwasher rack. Boil some white vinegar in your kettle as a useful way of removing hard water deposits.

WATER CONDITIONERS OR WATER SOFTENERS

Traditionally the water treatment market had one main solution to hard water. This solution was water softeners. However, in recent years alternative treatments have become increasingly popular, the most interesting of which is electromagnetic water conditioning.

Water Softeners work by ion exchange, so sodium replaces the calcium and/or magnesium in the water. Water

Conditioners on the other hand create a magnetic field around your pipework which alters the ions in the water so that they loose their ability to cause scale.

In tackling hard water, both methods will reduce limescale. Water Conditioners are significantly less expensive to start with and they have negligible running costs. Water Softeners cost a lot more but have the added effect that they will treat very small amounts of other metals such as Copper, Iron or Zinc.

Chapter 2

Occurrence, Properties and Uses of Compounds

NITROGEN

OCCURRENCE

Nitrogen occurs in free state in atmosphere, which contains 78% of nitrogen by volume. It is present in volcanic gases and gases evolved by burning coal. In combined state, nitrogen occurs in animals and plants as proteins, and as nitrates, such as $NaNO_3$ (chile salt peter), KNO_3 (nitre), nitrites and ammonium salts in the soil.

PREPARATION

Nitrogen can be easily prepared from its compounds.

By Heating Ammonium nitrite

Ammonium nitrite on heating decomposes to form nitrogen.

$$NH_4NO_2 \xrightarrow{\text{heat}} N_2 + 2H_2O$$

Since ammonium nitrite is very unstable, it cannot be stored and handled as such. Hence nitrogen is usually prepared by heating a mixture of ammonium chloride and sodium nitrite.

Laboratory Preparation of Nitrogen

In laboratory, dinitrogen is prepared by heating an

aqueous solution containing equivalent amounts of ammonium chloride and sodium nitrite.

$$NH_4CI + NaNO_2 \longrightarrow NH_2NO_2 + NaCI$$

$$NH_4NO_2 \xrightarrow{heat} N_2 + 2H_2O$$

During the preparation, small amounts of NO, ammonia and HNO_3 are also formed. The gas is freed from these by passing through a solution of potassium dichromate acidified with sulphuric acid and collected over water.

METHODS

By Heating Ammonium Dichromate

Ammonium dichromate on heating decomposes to give nitrogen gas.

$$(NH_4)_2 Cr_2O_7 \longrightarrow N_2 + 4H_2O + Cr_2O_3$$

This reaction is very vigorous and inconvenient for the preparation of nitrogen.

By the Oxidation of Ammonia

When ammonia vapours are passed over heated cupric oxide (CuO), nitrogen is formed.

$$2NH_3 + 3CuO \xrightarrow{heat} N_{2(g)} + 3Cu + 3H_2O$$

By the Action of Nitrous Acid on Urea

Urea when treated with nitrous acid gives nitrogen gas:

$$\underset{\text{Urea}}{NH_2CONH_2 9} + 2HNO_2 \longrightarrow 2N_2 + CO_2 + 3H_2O$$

Nitrous acid is unstable and is always prepared in the solution itself by the action of dilute HCl on sodium nitrite.

$$NaNO_2 + HCI \longrightarrow NaCI + \underset{\text{Nitrous acid}}{NHO_2}$$

$$NH_2CONH_2 + 2HNO_2 \longrightarrow 2N_2 + CO_2 + 3H_2O$$

The gas is bubbled through NaOH solution to remove CO_2 and then collected over water.

From Air

On commercial scale, nitrogen is obtained by the fractional distillation of liquefied air. The compressed carbon dioxide free air is liquefied by allowing it to expand in a region of low pressure.

The liquefied air is then subjected to fractional distillation. Nitrogen having lower boiling point (78 K) distils over first leaving behind oxygen. Nitrogen obtained from air contains traces of oxygen.

To remove these, the nitrogen so obtained is passed over heated copper turnings when copper reacts with oxygen to form CuO.

Physical Properties of Nitrogen

- Nitrogen gas (dinitrogen) is a colorless, odorless and tasteless gas.
- Nitrogen gas is slightly lighter than air.
- Nitrogen gas is very sparingly soluble (almost insoluble) in water: 23 mL per litre of water at 0°C and 1 atm pressure.
- Nitrogen gas liquefies at 77.2 K (–195.9°C) and solidifies at 63.2 K (209.9°C).
- It is not poisonous, but animals die in nitrogen atmosphere for want of oxygen.
- Naturally Occurring nitrogen consists of two istopes $^{14}_{7}N$ and $^{15}_{7}N$.
- $^{15}_{7}N$ is often used as an isotopic tracer.
- Nitrogen is neither combustible nor a supporter of combustion.

CHEMICAL PROPERTIES OF NITROGEN

Reactivity of Nitrogen

Nitrogen gas (dinitrogen) is chemically non-reactive at ordinary temperature due to:

- Its small atomic size.
- Its high bond order.

The N-N bond in the N_2 molecule has a formal triple bond character.

$$:N::N:N \equiv N \quad \text{Bond Length}: 109.8\,\text{pm}$$

The $N \equiv N$ bond distance is 109.8 pm and is very stable. Its bond dissociation energy is 946kJ mol^{-1}.

$$N_2(g) + 946kJ \rightarrow 2N_{(g)}$$

The dissociation constant is extremely small and even at 3000 K there is no appreciable dissociation. This indicates the strength of the N-N bond.

Reaction with Highly Electropositive Metals

Nitrogen readily combines with highly electropositive metals (Group 1), even at room temperature, to give nitrides.

For example,

$$6Li + N_2 \rightarrow 2Li_3N$$

These nitrides are ionic compounds in which nitrogen is present as a trinegative anion, N^{3-}. Thus, lithium nitride may be represented as $(Li^+)_3N^{3-}$. Ionic nitrides are crystalline compounds and have high melting points.

Reaction with Less Electropositive Metals

Nitrogen reacts with less electropositive such as magnesium, calcium etc., at higher temperatures, with boron and aluminium at red heat, and with silicon at white heat (very high temperature) to form the corresponding nitrides.

$$3Mg + N_{2(g)} \xrightarrow{\text{heat}} Mg_3N_2$$

$$2AI + N_{2(g)} \xrightarrow{\text{red heat}} 2AIN$$

$$2Si + 2N_{2(g)} \xrightarrow{\text{White heat}} Si_3N_4$$

The nitrides of alkaline Earth metals (group 2) are ionic, while those of groups 13 and 14 are covalent in nature.

Reaction with Transition Metals

Transition metals such as iron, manganese form nitrides

when heated with nitrogen at very high temperatures. The nitrides of the transition metals are interstitial compounds in which nitrogen atoms occupy the interstices (due to its small size). These nitrides are extremely hard and have high melting points.

Reactions with Non-metals

- Nitrogen reacts with many non-metals under suitable conditions.

  For example with hydrogen:

$$N_{2(g)} + 3H_{2(g)} \xrightarrow[750K,\,Pressure]{Fe,Mo} 2NH_{3(g)} \Delta H = -92kJ$$

- With oxygen, it forms nitric oxide, which being unstable reacts with more oxygen to form nitrogen dioxide.

$$N_{2(g)} + O_{2(g)} \xrightarrow{3300\,K} 2NO_{(g)}\ \Delta H = +181kJ$$

$$2N + O_2 \longrightarrow 2NO_2$$

This reaction between N_2 and O_2 also takes place during lightening.

Combination with Compounds

Nitrogen combines with certain compounds on strong heating.

For example,

With calcium carbide:

$$CaC_2(s) + N_{2(g)} \xrightarrow{13000\,K} \underset{\text{Calcium cyanide}}{CaCN_2} + C$$

With alumina in presence of carbon:

$$Al_2O_{3(s)} + N_{2(g)} + 3C_{(s)} \xrightarrow{2100\,K} \underset{\text{alu min um Nitride}}{2AIN} + 3CO$$

Calcium cyanamide, aluminium nitride and magnesium nitride hydrolyze on boiling with water to give ammonia.

$$Mg_3N_2 + 6H_2O \longrightarrow 3Mg(OH)_2 + 2NH_3$$

$$AIN + 3H_2O \longrightarrow Al(OH)_3 + NH_3$$

$$CaNCN + 3N_2O \longrightarrow CaCO_3 + 2NH_3$$

Hydrogen Bonding

Nitrogen in its compounds with hydrogen shows hydrogen bonding though to a lesser extent than oxygen. For example, ammonia (NH_3) shows intermolecular hydrogen bonding.

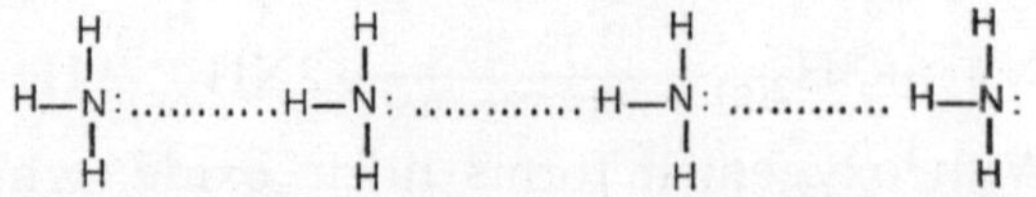

Catenation

The tendency to establish self-linkage of the type M-M-M- leading to the formation of a chain or ring is called catenation. Due to p - p overlap nitrogen forms multiple bonds with itself as well as with carbon and oxygen e.g.,

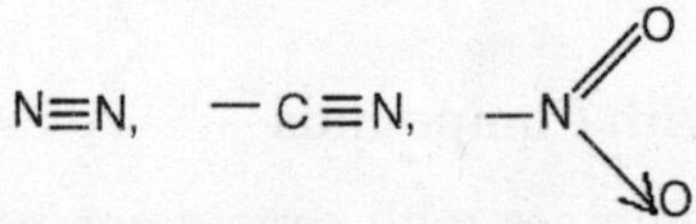

Thus, nitrogen shows catenation tendency though weaker than that shown by carbon. The following compounds show the catenation property of nitrogen.

$R - N = N - NR_2$ $R_2N - N = N - NR_2$

$RN = N - N(R) - N = NR$

$RN = N - N(R) - N = N - N(R) - N = NR$

where R is some organic alkyl or aryl group. In some compounds R may also be H.

Uses

- Nitrogen gas is mainly used for the manufacture of compounds such as, ammonia, nitric acid, calcium cyanamide etc.
- Nitrogen gas is used for providing inert atmosphere in many metallurgical operations.
- Liquid nitrogen is used for producing low tempera-

tures required in cryosurgery, preservation of biological materials etc.

FIXATION

Nitrogen is present in the elemental state in the atmosphere to the extent of about 78% by volume. Nitrogen is important for life processes in all living organisms. But, free nitrogen cannot be directly assimilated by the living cells. Therefore, it is desirable to convert the free atmospheric nitrogen into useful nitrogen compounds. The process of conversion of free atmospheric nitrogen into useful nitrogen compounds is called fixation of nitrogen or nitrogen fixation.

Fixation of Atmospheric Nitrogen by Electric Discharge

During rains, when lightning strikes, nitrogen and oxygen of the atmosphere combine to form nitric oxide. This is further oxidized to nitrogen dioxide in the presence of excess of air. Nitrogen dioxide combines with rainwater in presence of oxygen to produce nitric acid.

Nitric acid reacts with some alkalies in the soil, e.g., limestone ($CaCO_3$), to form nitrates. These nitrates serve as plant food and are responsible for their growth.

$$\underset{\text{atimosphere}}{N_{2(g)} + O_{2(g)}} \xrightarrow{\text{Electric Discharge}} \underset{\text{Nitric Oxide}}{2NO_{(g)}}$$

$$2NO + O_2 \longrightarrow 2NO_2$$

$$4NO_2 + 2H_2O + O_2 \longrightarrow 4HNO_3$$

$$2HNO_3 + \underset{\text{limestone in the soil}}{CaCO_3} \longrightarrow \underset{\text{Acts as plant food}}{Ca(NO_3)_2} + H_2O + CO_2$$

Fixation of Atmospheric Nitrogen by Symbiotic Bacteria

Certain plants such as peas, beans, etc., have some special kind of nitrogen fixing bacteria (symbiotic bacteria) in the nodules of their roots.

These bacteria fix free nitrogen of the air as nitrogen compounds. These compounds spread into the soil after the decay of such plants.

Artificial Fixation of Nitrogen

Large amounts of free atmospheric nitrogen is however, fixed artificially by a number of methods. Three important methods are:

The Cyanamide Process

Nitrogen is also fixed as calcium cyanamide on heating it with calcium carbide at 1000°C in an electric furnace.

$$\underset{\text{Calcium Carbide}}{CaC_2} + N \xrightarrow{\text{Electric Furnace}} \underset{\text{Calcium Cyanamide}}{CaCN_2} + C$$

The mixture of calcium cyanamide and carbon (trade name nitrolim) is an important fertilizer. Calcium cyanamide is also be used as a source of ammonia. The ammonia so produced can be converted into useful fertilizers. Calcium cyanamide is decomposed by water to give ammonia.

$$CaCN_2 + 3N_2O \longrightarrow CaCO_3 + 2NH_3$$

$$2NH_3 + H_2SO_4 \longrightarrow (NH_4)_2SO_4$$

When a mixture of nitrogen (obtained from the liquefied air) and hydrogen in the ratio 1:3, under a pressure of 200 - 800 atmospheres is passed over iron and molybdenum catalyst at about 450°C, ammonia is produced.

$$N_2 + 3H_2 \xrightarrow[450°C]{Fe/Mo} 2NH_3$$

Ammonia is a basic raw material for preparing many salts and fertilizers. A part of ammonia produced from the fixation of nitrogen is used for the manufacture of nitric acid by Ostwald's process.

$$4NH_3 + 5O_2 \xrightarrow{Pt, 8000°C} 4NO + 6H_2°$$

$$2NO + O_2 \longrightarrow 2NO_2$$

$$4NO_2 + 2H_2O + O_2 \longrightarrow 4HNO_3$$

Ammonia and nitric acid so produced are used for the production of many ammonium salts and nitrates, which act as fertilizers, e.g., calcium ammonium nitrate (CAN).

When nitrogen and oxygen are made to combine by passing air through an electric arc (~ 3000°C), nitric oxide is formed. It combines with more oxygen to form nitrogen dioxide which when absorbed in water, produces nitric acid.

$$N_2 + O_2 \xrightarrow[\text{electric Arc}]{3000°C} 2NO$$

$$2NO + O_2 \longrightarrow 2NO_2$$

$$4NO_2 + 2H_2O + O_2 \longrightarrow 4HNO_3$$

Problems:

- How do you account for the inert character of dinitrogen?

Solution: The non-reactivity or inertness of nitrogen is due to:

Its small atomic size. Its high bond order. The N-N bond in the N_2 molecule has a formal triple bond character

$$:N::N:N \equiv N \text{ Bond Length} : 109.8 \text{ pm}$$

The $N \equiv N$ bond distance is 109.8 pm and is very stable. Its bond dissociation energy is 946kJ mol^{-1}.

$$N_{2(g)} + 946\,kJ \longrightarrow 2N_{(g)}$$

The dissociation constant is extremely small and even at 3000 K there is no appreciable dissociation. This indicates the strength of the N–N bond and accounts for its inert character.

- When a heap of ammonium dichromate is ignited, it erupts like a volcano, why?

Solution: When a heap of ammonium dichromate is ignited the chemical reaction that occurs is as follows:

$$(NH_4)_2Cr_2O_7 \longrightarrow N_2 + 4H_2O + Cr_2O_3 + \text{Heat}$$

Ammonium dichromate on heating decomposes to give nitrogen gas. But this reaction is so very vigorous that it is accompanied by the production of heat, light and lot of gases. As ammonium dichromate decomposes violently to form chromium oxide, this method is inconvenient for the preparation of nitrogen. The whole process looks like the eruption of a volcano and so this reaction is also referred to as chemical volcano.

AMMONIA

APPEARANCE

Ammonia is a colorless gas with a sharp, irritating odor. It is extremely soluble in water and is lighter than air.

PHYSICAL PROPERTIES OF NH_3

	Ammonia, NH_3
Atomic mass:	17.0306 g/mol
Melting point	–77.7 °C
Boiling point	–33.3 °C

HISTORY

Ammonia was first isolated by Priestley in 1774 with the aid of a pneumatic trough filled with mercury instead of water.

NATURAL ABUNDANCE

Ammonia is produced by the anaerobic decay of organic material. Ammonia was the first complex molecule to be identified in interstellar space and solid NH_3makes up the rings on Saturn.

INDUSTRIAL PRODUCTION

Prior to the 20th century, ammonia was made by destructive distillation of animal parts such as hoofs, horns, etc. The common name for ammonia was Spirits of Hartshorn. Haber worked out a synthetic route to catalytically produce ammonia from the elements at high pressure and temperature. The Haber process is still used today.

$$N_2(g) + 3\ H_2(g) \rightarrow 2\ NH_3(g) \quad \Delta H = -92\ kJ\ DS = 199\ J/K$$

INDUSTRIAL USES

Ammonia production is the largest consumer of nitrogen

and is the gateway compound to most other nitrogen compounds produced industrially. Enormous quantities of ammonia are produced worldwide.

Much of the ammonia produced in the USA is used directly for fertilizer. The figure below shows the various uses of ammonia. The majority of all ammonia and ammonium nitrate producted is for use as fertilizer.

Approximately 70% of all NH_3 produced is directly used for fertilizer. Farmers inject the pressurized liquid ammonia into the soil. Ammonia is so soluble in water and has such a high affinity for polar groups such as hydroxyl groups that the ammonia is almost instantly taken up by the soil and very little is lost to the atmosphere.

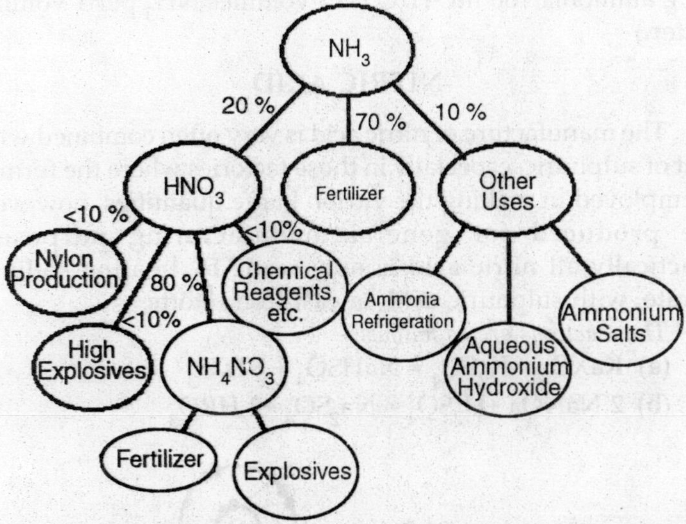

Other important uses of ammonia include the production of nitric acid, use as a refrigerant, and use as a non-aqueous solvent. Ammonias use as a refrigerant comes from the fact that it is a condensable liquid. Ammonia is currently commonly used as the refrigerant in large industrial refrigerators and freezers. Its main disadvantage in residential use stems from its toxicity should a leak in the system develop. Liquid ammonia is also the most common, widely used non-aqueous solvent. Its boiling point, -33.3 °C, requires that vessels be refrigerated or pressurized. Ammonia is used in the manufacture of numerous other chemicals and products ranging from dyes to plastics.

GAS DENSITY OF NH_3

The density of NH_3 is 0.6826 g/L at 25 °C and 1 atm. This is about half that of airs, thus it tends to rise in a still room.

GAS SOLUBILITY OF NH_3

Ammonia is extremely soluble in water. As much as 89.9 g dissolve per L at 0 °C. This means that 1 mL of water will dissolve 1183 mL ammonia! At 100 °C, the solubility drops to 7.4 g ammonia/100 mL H_2O (113 volumes NH_3 per 1 volume water.)

NITRIC ACID

The manufacture of nitric acid is very often combined with that of sulphuric, especially in those factories where the former is employed in making the vitriol. Large quantities, however, are produced for general manufacturing purposes. Practically all nitric acid is now made by treating sodium nitrate, with sulphuric acid, in cast-iron retorts.

The reactions are as follows:

(a) $KaXO_3 + H_2SO_4 = NaHSO_4 + I\text{-}IKO_3$

(b) $2\ NaNOa + H_2SO_4 = Na_2SO_4 + 2\ HRO_3$

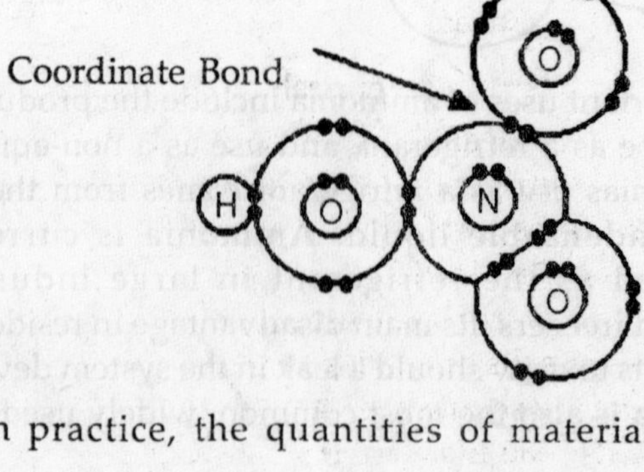

In practice, the quantities of material used do not

correspond with either of these equations, but the charge is so regulated that a mixture of acid and neutral sulphates of sodium, which remains liquid at the temperature employed, is left in the retort.

If reaction:

- Were followed, too much sulphuric acid would be used for proHtable working, except in soda works, where the resulting acid sulphate might be used in the salt-cake furnace.
- Is carried out, the temperature must be very high, the neutral sulphate solidifies in the retort and is difficult to remove, and the resulting nitric acid may be partly decomposed by the heat before it can escape from the retort, thus causing a diminished yield and a product discolored by the oxides of nitrogen produced.

The sulphuric acid employed is usually that from the lead pan evaporation, oil of vitriol of 66° Be. is used, although this decomposes part of the nitric acid formed. The sodium nitrate used is the purified Cl~ili saltpetre, containing from 98 to 99 per cent of NaNOs, when dried. It should be free from sodium chloride to avoid contaminating the nitric acid with hydrochloric acid. The size of the charge depends on the capacity of the plant, but in some more modern factories it amounts to as much as 1200 pounds of nitrate, with somewhat more than an equal weight of sulphuric acid.

'1'he-old style of plant consisted of a horizontal cast-iron cylinder, 5 or 6 feet long, set over a fireplace in such a way that the flames played over the sides and top, heating all parts to a high temperature. Another and better form of plant, in which the cast-iron retort (A) is entirely surrounded by the flames from the grate. Cast iron is but little attacked by concentrated nitric acid or its vapors, and it is important to keep the retort hot enough in all parts, to prevent condensation of the acid.

In the lower part of the retort is a pipe, by which the melte!l residue of "nitre cake" is run off, after the reaction is finished. For condensing the acid vapors which escape from the retort, a series of glass or earthen ware 'Woulfe bottles (bombonnes) are employed.

The first two or three of these bottles are generally placed over the flue by which the fire gases pass to the chimney. Being thus warmed, there is less danger of breakage by the high heat of the vapors from the retort.

At the end of the series is usually placed a coke towel', fed with water or concentrated sulphuric acid, to condense the fumes escaping fro the bombonnes.

Usually, no water is admitted to the bombonnes unless a dilute acid is required; but they are sometimes placed en cascade, to allow the condensed acid to flow through the series in a direction opposite to the movement of the acid vapors.

The most concentrated acid is condensed in the first two or three bomboHnes, but is contaminated with sulphuric acid and nitrogen oxides. The last of the series contains dilute acid, which is Contaminated with chlorine. In the middle bottles is a pure acid of moderate strength. Owing to more or less reduction of the nitric acid in the retort, the condensed acid has a yellow or red colour, due to the absorbed nitrous vapors.

These are undesirable in a commercial acid, and must be removed by "bleaching "; the acid is heated to about DO° C., and warm ail' blown in, which cal'l'ics away the nitrogen oxides, and is then passed through the coke tower for their recovery. Guttmann's apparatus is more modern.

The large cast-iron retort is made in three pieces and is entirely surrounded by the flames from the grate. The gases from the retort pass into a system of vertical earthenware pipes (AA), having very thin walls and joined at the top by 1800 bends, while they open at the bottom into a nearly horizontal collecting pipe (B B), which is divided into chambers by diaphragms.

These chambers are joined by U-tubes, passing under the diaphragms. The diaphragms force the acid vapors to pass up one pipe and down of the vertical pipes allow very efficient cooling by exposure to the air alone, or they may be placed in a tank of cold water. By this very rapid cooling, the acid vapors are condensed quickly.

Hot air at 80° C. is injected from (F) into the outlet: pipe (D), where it converts some of the nitrous vapors to nitric acid, increasing the yield materially. (The uncondensed nitrous vapors pass into the Lunge-Rohrmann plate tower (E), where the nitrogen oxides are absorbed in sulphuric acid or water.) If the vapors remain in the retort too long, part of the acid is decomposed, and nitrogen peroxide is formed, and absorbed by the condensed acid, to which it imparts a red colour.

But since there is a good draught through the apparatus, the vapors are drawn out of the retort very soon after they are evolved, and are at once condensed. Thus very little peroxide is formed, and a light colored, concentrated acid is obtained directly. It is claimed that acid of 40° Be. requiring no "bleaching," may be thus made, and that, with water-cooled pipes, 98 per cent of the theoretical yield is obtained as concentrated acid, while 2 per cent condenses in the Lunge-Rohrmann tower.

Hart's tube condenser, for nitric acid, is a new form of apparatus. It is made of glass and earthenware tubes, and is placed above the brick arch covering the retort, thus occupying but little floor space.

The vapour from the retort I III A (A) passes into the pot, and thence through the vertical earthenware tube (C). From (C) to (D) extend a number of glass tubes, are slightly inclined towards (C), and which are cooled by jets of water from the perforated pipe (EE). From (D), the uncondensed vapors pass to a Lunge tower or a coke tower. The acid condensed in the glass tubes flows back into (C), thus coming into contact with the hot vapors from the retort. This heats the acid so hot, that all the the U-tube (F) into bottles or carboys. In this apparatus, the acid is condensed very quickly and but little nitrogen peroxide is formed.

If frothing occurs in the retort, the overflow is caught in the pot (B), and may be removed without difficulty. The flow of water from (E) can be so regulated that all or nearly all of it is evaporated on the surface of the glass tubes, thus securing the greatest cooling effect, with small consumption of water. Any condenser water not evaporated is caught in the trough

(G). The chief repairs are of broken tubes, which are cheaply and easily replaced. The strength of the nitric acid produced in any apparatus depends upon the strength of the sulphuric acid, on the temperature of the retort, and on the purity of the sodium nitrate. "With sulphuric acid, the nitric acid varies (40° to 42° Be.). If the sodium nitrate contains chlorides, some of the nitric acid is decomposed by the hydrochloric acid produced; thus:

$$HNO_3 + HCl = H_2O + NO_2 + Cl.$$

It is claimed that with Guttmann's apparatus, an acid of 1.5 to 1.52. (50° Be.), containing about 95 per cent HXO_3, can be made. For chemically pure acid, perfectly pure materials should be used, although formerly the common acid was purified by treating with silver and barium nitrates, and redistilling. But concentrated acid cannot be distilled without some decomposition, and the product must be"bleached" by heating and blowing in pure air.

Fuming nitric acid is a solution of nitrogen peroxide in concentrated nitric acid. It is red in colour and has a specific gravity of 1.55 to 1.62.

To make this, perfectly dry sodium nitrate amI oil of vitriol are used. The reaction is carried so far that neutral sulphate of sodium is formed by the action of the acid sulphate on the nitrate:

$$2{:}NaNO_3 + 2\,NaHSO_4 = 2Na_2SO_4 + 2NO_2 + H_2O + O.$$

The nitrogen peroxide formed dissolves in the nitric acid to form the fuming acid. A little powdered starch is sometimes added to assist in the reduction of"~henitric acid. An impure fuming acid is sometimes prepared by distilling a mixture of concentrated nitric and sulphuric acids. Nitric acid is largely used in the manufacture of explosives; for parting gold and silver; in the manufacture of coal-tar dyes; as a various metallic nitrates. The pure acid also decomposes on exposure to strong light and becomes yellow (XO_2).

FEATURES

Nitric acid (HNO_3) is a colorless, liquid acid widely used in the manufacturing of explosives and fertilizers. When dissolved in water, molecules of nitric acid separate (or dissociate) into hydrogen ions (H^+) and nitrate ions (NO_3). The fact that nearly every nitric acid molecule dissociates is what makes nitric acid a strong acid. Nitric acid is often the starting material in the industrial production of nitrates for fertilizers.

Plants take up nitrogen from the soil in the form of ammonium ions ($NH_4{}^+$) and nitrate ions, and along with carbon containing molecules made during photosynthesis, these ions are used to synthesize amino acids, from which proteins are made.

Within the past hundred years the demand for nitrogen fertilizers has grown dramatically as the need for fertilizers for agriculture has grown. The natural manner in which nitrates reach the soil involves the reaction of nitrogen gas and oxygen gas in the atmosphere to form nitrogen dioxide gas (NO_2), which then reacts with atmospheric water, making nitric acid, which provides a natural source of nitrates in water and soil.

During World War I, the Germans were very interested in using ammonium nitrate (NH_4NO_3), a salt of nitric acid, as an explosive. Many organic nitrates such as nitroglycerine and TNT are also highly explosive. Nitric acid is made by the reaction of ammonia with oxygen gas. The nitric acid which is produced can then be used to make a variety of compounds. This is a process that has allowed large amounts of fertilizers to be produced relatively inexpensively.

Nitric acid is also formed from the reaction of nitrogen oxides produced during the combustion (burning) of fossil fuels in automobileengines. These nitrogen oxides react with water in the atmosphere and form nitric acid, one cause of acid rain. High levels of nitrates in drinking water can contribute to the formation of nitrosamines, a group of carcinogenic (cancer causing) compounds.

A clear, colorless liquid that is one of the strongest mineral acids. In air, nitric acid (HNO_3) fumes and turns yellow due to its decomposition into nitrogen dioxide (NO_2). Nitric acid is highly corrosive and a powerful oxidizing agent. It attacks most metals, resulting in the formation of nitrates. With many organic compounds,nitration occurs. Melting point –42°C, boiling point 83°C. Nitric acid is used in the production of fertilizers, explosives, plastics, dyes, and rocket propellants. It was called *aqua fortis* ("strong water") by the alchemists, a name by which it continued to be known commercially.

PREPARATION

Nitric acid can be prepared in the lab by the distillation of a mixture of an alkali-metal nitrate and concentratedsulfuric acid. Industrially, it is produced by the Ostwald process (named after the Latvian-born German physical chemist Friedrich Wilhelm Ostwald who developed).

In this process, ammonia is oxidized by air to nitric oxide, at 900°C and 1–8 atm using a platinum or platinum-rhodium catalyst in the form of a thin wire gauze. The nitric oxide is further oxidized to nitrogen dioxide, which is dissolved in water to give 60% nitric acid.

USES

Nitric acid was commonly used as an oxidizer in liquid-propellant rocket engines. It most often took the form of RFNA (red fuming nitric acid), containing 5–20% dissolved nitrogen dioxide.

Compared to concentrated nitric acid (also known as white fuming nitric acid), RFNA is more energetic and more stable to store but produces poisonous red-brown fumes. Because nitric acid is normally highly corrosive it can only be stored and piped by a few materials such as stainless steel.

However, the addition of a small concentration of fluoride ions inhibits the corrosive action and gives a form known as IRFNA (inhibited red fuming nitric acid).

Like nitrogen tetroxide, it is hypergolic (reacts upon contact with) hydrazine, MMH(monomethyl hydrazine), and UDMH (unsymmetrical dimethyl hydrazine).

EFFECTS

Concentrated nitric acid is capable of producing severe

burns of the skin. Swallowing the acid leads to intense burning pain and ulceration of the mouth and throat. Treatment is by immediate administration of alkaline solutions, followed by milk or olive oil.

CARBON

OVERVIEW

Carbon is a set of APIs for developing full-featured, high-performance, and reliable applications for Mac OS X. Carbon enables C, Objective-C and C++ developers to take advantage of Mac OS X-specific features, including an advanced user interface toolkit, an efficient event-handling mechanism, the Quartz 2D graphics library, and multiprocessing support.

In addition, other industry-standard C and C++ APIs are easily available to Carbon developers, providing access to such services as the OpenGL drawing system, the Mach micro-kernel, and BSD operating-system services, and even the Cocoa framework. And you can use virtually any Macintosh development environment that supports Intel and PowerPC to build Carbon applications.

If you are already familiar with a C/C++ API, even if on another platform, Carbon simplifies porting your application to Mac OS X. Carbon provides access to the advanced features in the Macintosh operating system and hardware, and is continually being updated to support new Mac OS X technologies, so that you can upgrade your applications in step with innovative features as Apple releases new APIs.

The Carbon APIs are also well-suited to cross-platform development. Major software vendors use Carbon to build some of the most popular commercial Macintosh applications.

You can use Apple's own IDE Xcode and Interface Builder, specifically designed for constructing applications on Mac OS X. Or you can use any of the third-party tools that support Intel and PowerPC to build Carbon applications. New applications should use Mach-O, Mac OS X's native executable format. The Carbon framework contains thousands of functions, data structures, and constants. Related functions and data structures are organized into functional groups, usually referred to as managers or services.

The managers include specific APIs for managing user interfaces, file access, text editing, and threads, to name a few. For example, the File Manager provides an API for application access to the file system, and Navigation Services provides an API for helping users navigate the file system.

Apple's Carbon is the essential building block for writing C, Objective-C and C++ applications that run on Mac OS X. You gain all the benefits of Mac OS X while using some of the simplest yet most capable C interfaces available. This combination makes it easier than ever to create world-class applications for this world-class operating system.

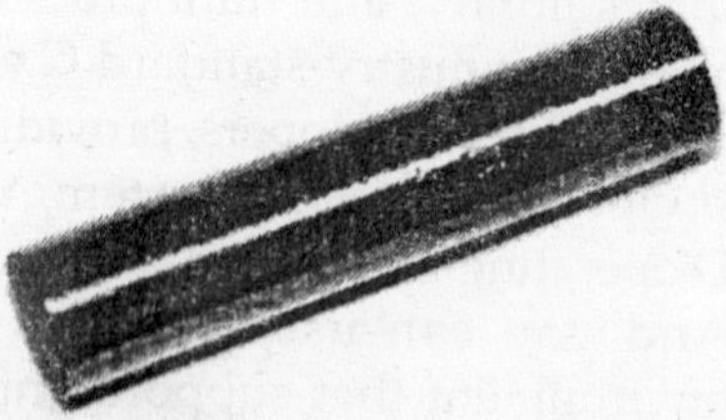

An advanced material of pure carbon combining glassy andceramic properties with those of graphite. Unlike graphite, vitreous carbon has a fullerene-related microstructure. Image copyright: smart-elements.com

Carbon is the non-metallic chemical element of atomic number 6 in Group 14 of the periodic table, symbol C, atomic weight 12.01, specific gravity as graphite 2.25, as diamond 3.51. Its stable isotopes are ^{12}C (98.90%) and ^{13}C (1.10%). The weight of the ^{12}C atom is the international standard on which atomic weights are based. It is defined as weighing exactly 12.00000 atomic mass units. Carbon has been known since prehistoric times. It gets its name from *carbo*, the Latin word for charcoal, which is almost pure carbon. In various forms, carbon is found not only on Earth, but in the atmospheres of other planets, in the Sun and stars, in comets, and in some meteorites.

On Earth, carbon can be considered to be the most important of all the chemical elements, because it is the essential element in practically all of the chemical compounds in living things.

Carbon compounds are what make the processes of life work. Beyond Earth, carbon-atom nuclei are an essential part of the nuclear fusion reactions that produce the energy of the Sun and of many other stars. Without carbon, the Sun would be cold and dark.

It is obviously impossible to summarize the properties of carbon's millions of compounds in one place. Introductory textbooks of organic chemistry generally run well over a thousand pages, and biochemistry textbooks run thousands more. But organic compounds can be classified into families that have similar properties, because they have certain groupings of atoms in common.

HISTORY

Latin. carbo: Charcoal carbon, an element of prehistoric discovery, is very widely distributed in nature. It is found in abundance in the sun, stars, comets, and atmospheres of most planets. Carbon in the form of microscopic diamonds is found in some meteorites.

Natural diamonds are found in kimberlite of ancient volcanic "pipes," found in South Africa, Arkansas, and elsewhere. Diamonds are now also being recovered from the ocean floor off the Cape of Good Hope. About 30% of all industrial diamonds used in the U.S. are now made synthetically. The energy of the sun and stars can be attributed at least in part to the well-known carbon-nitrogen cycle.

FORMS

Carbon is found free in nature in three allotropic forms: graphite, diamond, and fullerines. A fourth form, known as "white" carbon, is now thought to exist. Ceraphite is one of the softest known materials while diamond is one of the hardest.

Graphite exists in two forms: alpha and beta. These have identical physical properties, except for their crystal structure. Naturally occurring graphites are reported to contain as much as 30% of the rhombohedral (beta) form, whereas synthetic materials contain only the alpha form.

The hexagonal alpha type can be converted to the beta by mechanical treatment, and the beta form reverts to the alpha on heating it above 1000°C.

A new allotropic form of carbon was produced during the sublimation of pyrolytic graphite at low pressures. Under free-vaporization conditions above ~2550°K, "white" carbon forms as small transparent crystals on the edges of the planes of graphite. The interplanar spacings of "white" carbon are identical to those of carbon form noted in the graphite gneiss from the Ries (meteroritic) Crater of Germany. "White" carbon is a transparent birefringent material. Little information is presently available about this allotrope.

COMPOUNDS

In combination, carbon is found as carbon dioxide in the atmosphere of the earth and dissolved in all natural waters. It is a component of great rock masses in the form of carbonates of calcium (limestone), magnesium, and iron. Coal, petroleum, and natural gas are chiefly hydrocarbons.

Carbon is unique among the elements in the vast number and variety of compounds it can form. With hydrogen, oxygen, nitrogen, and other elements, it forms a very large number of compounds, carbon atom often being linked to another carbon atom. There are close to ten million known carbon compounds, many thousands of which are vital to organic and life processes.

Without carbon, the basis for life would be impossible. While it has been thought that silicon might take the place of carbon in forming a host of similar compounds, it is now not possible to form stable compounds with very long chains of silicon atoms. The atmosphere of Mars contains 96.2% CO_2. Some of the most important compounds of carbon are carbon dioxide (CO_2), carbon monoxide (CO), carbon disulfide (CS_2), chloroform ($CHCl_3$), carbon tetrachloride (CCl_4), methane (CH_4), ethylene (C_2H_4), acetylene (C_2H_2), benzene (C_6H_6), acetic acid (CH_3COOH), and their derivatives.

ISOTOPES

Carbon has seven isotopes. International Union of Pure and Applied Chemistry adopted the isotope carbon-12 as the basis for atomic weights. Carbon-14, an isotope with a half-life of 5715 years, has been widely used to date such materials as wood, archaeological specimens, etc.

CARBONDIOXIDE

Chemical compound composed of one carbon and two oxygen atoms joined together by double covalent bonds. Carbon dioxide is a minor component of Earth's atmosphere (0.03 per cent by volume) but one that is essential to terrestrial life. It is the major component of the thick atmosphere of Venus and of the thin atmosphere of Mars, and is an important greenhouse gas. Below -78°C it forms a solid, commonly known as dry ice, and is found in this state, together with water ice, in the Martian polar caps and in comets.

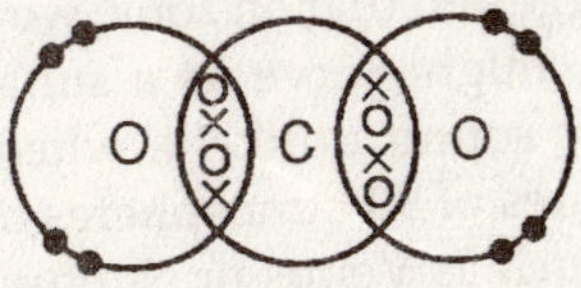

Fig. Carbon Dioxide Molecule

Carbon dioxide is taken in by plants during photosynthesis and produced by animal respiration. The building of an organism requires a way of making organic molecules from carbon dioxide.

This is achieved during the Calvin cycle of photosynthesis in green plants. By means of this cycle, plants incorporate carbon dioxide intosugars which form the basis for all of the organic molecules synthesized by cells. Carbon dioxide is also an important greenhouse gas.

Carbon dioxide is prepared in the laboratory by reacting a carbonate with acid; industrially it is obtained by calcining limestone, burning coke in excess air, or from fermentation.

Liquid carbon dioxide, formed under pressure, is used in fire extinguishers. Carbon dioxide is also used to make carbonated drinks. When dissolved in water an equilibrium is set up, with carbonate, bicarbonate, and hydrogen ions formed, and a low concentration of carbonic acid (H_2CO_3).

Density (at 0°C)	1.98 g/dm^3
Melting point	–56.6°C
Boiling point	–78.5°C
Molar mass	44.0095 g/mol

DRY ICE

Solid carbon dioxide, or dry ice, is used as a refrigerant for transporting perishables. It is made by compressing carbon dioxide gas to a pressure of about – 7MN/m^2 at –57°C, when it liquefies; it is then expanded adiabatically to atmospheric pressure and cools, solid carbon dioxide "snow" separating. This is compressed into blocks. Dry ice sublimes at -78.5°C;.

SOLVENT

It has been suggested that on some worlds carbon dioxide, rather than water, might serve as a suitable solventfor life. Although this may sound unlikely, when carbon dioxide is subjected to pressures of 90 atmospheres or more, such as are found on planets such as Venus or Neptune, it enters a quasi-liquid, supercritical state. In this unfamiliar form, carbon dioxide is as heavy as water but almost as easy to move through as air because molecules of carbon dioxide don't stick together as tightly as the molecules in most liquids. When certainenzymes (biological catalysts) are placed in supercritical carbon dioxide, they work as well as they do in other substances that have been suggested as alternative solvents for life, such as hexane and ether.

PHOSPHORIC ACID

OCCURRENCE

About 10 million tons of phosphoric acid, H_3PO_4, are produced in this country each year. Most of the acid (about 80%) is used in the production of agricultural fertilizers, with the remainder being used for detergent additives (about 10%), cleaners, insecticide production, and cattle feed additives. The commercial method of preparation is the addition of sulfuric acid to phosphate rock.

$$3\ H_2SO_4(l) + Ca_3(PO_4)_2(s) + 6\ H_2O(l) \rightarrow {}_2\ H_3PO_4(s) + 3\ CaSO_4.2\ H_2O(s)$$

Pure anhydrous phosphoric acid is a white solid which melts at 42.35EC to form a viscous liquid. In aqueous solution, phosphoric acid behaves as a triprotic acid, having three ionizable hydrogen atoms. The hydrogen ions are lost sequentially.

$H_3PO_4(aq) \leftrightarrows H+(aq) + H_2PO_4\ K_{a1} = 7.5 \times 10^{-3}$

$H_2PO_4-(aq) \leftrightarrows H+(aq) + HPO_4\ K_{a2} = 6.2 \times 10^{-8}$

$HPO_4{}^{2-}(aq) \leftrightarrows H+(aq) + PO_4{}^{3-}(aq)\ K_{a3} = 1.7 \times 10^{-12}$

Phosphoric acid is not a particularly strong acid as indicated by its first dissociation constant. It is a stronger acid than acetic acid, but weaker than sulfuric acid and hydrochloric acid. Each successive dissociation step occurs with decreasing ease. Thus, the ion H_2PO_4G is a very weak acid, and $HPO_4{}^{2-}$ is an extremely weak acid.

PREPARATION

Salts of phosphoric acid can be formed by replacing one, two or three of the hydrogen ions. For example, NaH_2PO_4, sodium dehydrogenate phosphate, can be formed by reacting one mole of phosphoric acid with one mole of sodium hydroxide.

$$H_3PO_4(aq) + NaOH(aq) \rightarrow NaH_2PO_4(aq) + H_2O(R)$$

[*net ionic form*: $H_3PO_4(aq) + OHG(aq) \rightarrow H_2PO_4G(aq) + H_2O(R)$]

Similarly, Na_2HPO_4 (disodium hydrogen phosphate) and Na_3PO_4, (trisodium phosphate) could be formed by the reaction of one mole of H_3PO_4 with two and three moles of NaOH, respectively. Salts containing the anion H_2PO_4G are weakly acidic. The tendency of this ion to dissociate is greater than its tendency to hydrolyse, that is, its Ka_2, is larger than its Kb.

$$H_2PO_4^-(aq) \leftrightarrows H+(aq) + HPO_4^{2-}(aq)\ K_{a2} = 6.2 \times 10^{-8}$$

$$H_2PO_4- (aq) + H_2O(l) \leftrightarrows H_3PO_4(aq) + OH^-(aq)\ K_b = K_w/K_{a1} = 1.3 \times 10^{-12}$$

Because $H_2PO_4^-$ is weakly acidic and of low toxicity, it is used as the acid in some baking powders. These baking powders contain NaH_2PO_4 and $NaHCO_3$ (sodium bicarbonate).

The leavening action of baking powders results from the production of carbon dioxide gas by an acid-base reaction between these two ingredients.

$$H_2PO_4^-(aq) + HCO_3^-(aq) \rightarrow HPO_4^{2-}(aq) + H_2O(l) + CO_2(g)$$

In the reaction between them, $H_2PO_4^-$ acts as the Brønsted-Lowry acid, HCO_3^- as the base. A comparison of the ionization constants for these two ions reveals that $H_2PO_4^-$ is a stronger acid than HCO_3^-.

$$H_2PO_4^- (aq) \leftrightarrows H+(aq) + HPO_4^{2-}(aq)\ K_{a2} = 6.2 \times 10^{-8}$$

$$HCO_3^- (aq) \leftrightarrows H+(aq) + CO_3^{2-}(aq)\ Ka_2 = 4.8 \times 10^{-11}$$

Salts containing the anion HPO_4^- are weakly basic. The tendency of this ion to hydrolyses is greater than its tendency to dissociate.

$$HPO_4^{2-}(aq) \leftrightarrows H+(aq) + PO_4^{3-}(aq)\ K_{a3} = 1.7 \times 10^{-12}$$

$$HPO_4^{2-}(aq) + H_2O(l) \leftrightarrows H_2PO_4^-(aq) + OH^-(aq)\ K_b = Kw/K_{a2} = 1.6 \times 10^{-7}$$

Solutions containing the phosphate ion, PO_4^{3-}, are quite basic. This ion has no acidic hydrogen, and its base ionization constant (hydrolysis constant) is relatively large.

$$PO_4^{3-}(aq) + H_2O(aq) \leftrightarrows HPO_4^{2-}(aq) + OH^-(aq)\ K_b = 5.9 \times 10^{-3}$$

As a result, solutions of soluble phosphates tend to have the same slippery, soapy feel as solutions of strong bases, such as NaOH or KOH. Phosphoric acid is used primarily in the manufacture of fertilizers, detergents, and pharmaceuticals. In the steel industry, it is used to clean and rust-proof the product.

It is also used as a flavoring agent in carbonated beverages (read the ingredients list on a can of Coca-Cola), beer, jams, jellies and cheeses. In foods, phosphoric acid provides a tart, acidic flavour.

A recent study reported in the journal *Epidemiology*, found that drinking two or more cola beverages per day doubled the risk of chronic kidney disease.

Cola beverages have been associated with kidney changes that promote kidney stones, which may be a result of the phosphoric acid in colas.

In the manufacture of detergents, phosphoric acid is used to produce water softeners. Water softeners remove Ca^{2+} and Mg^{2+} ions from hard water. If not removed, these hard-water ions react with soap and form insoluble deposits that cling to laundry and the washing machine. Phosphates produced from phosphoric acid are used extensively as water softeners (builders) in detergents.

The most widely used phosphorus compound in solid detergent mixtures is sodium tripolyphosphate, $Na_5P_3O_{10}$. As a water softener, sodium tripolyphosphate binds to Ca^{2+} and Mg^{2+}, forming soluble chemical species, called complexes or chelates. These complexes prevent the Ca^{2+} and Mg^{2+} from reacting with soap and forming deposits.

Most phosphoric acid is used in the production of fertilizers. Phosphorus is one of the elements essential for plant growth. Organic phosphates are the compounds which provide the energy for most of the chemical reactions that occur in living cells. Therefore, enriching soils with phosphate fertilizers enhances plant growth.

Increasing the phosphate concentration in surface waters also enhances the growth of aquatic plant life. Run-off from fertilized farm lands can stimulate plant growth in lakes and streams.

Waste water that contains phosphates from detergents can have the same effect. Lakes that are rich in plant nutrients suffer from accelerated eutrophication. When the lush aquatic plant growth in a nutrient-rich lake dies, the decomposition of the dead plant material consumes dissolved oxygen. This consumption reduces the level of dissolved oxygen to a point where it is insufficient to support animal life.

To reduce the threat of lake eutrophication, many localities have banned the use of phosphates in detergents. In some cases, the phosphates have been replaced by carbonates. In others, new detergents have been developed that do not react with the Ca^{2+} and Mg^{2+} ions of hard water.

SULPHUR DIOXIDE

OCCURRENCE

Sulphur dioxide is a yellowish, acrid gas. It is highly water-soluble and has a typical odour of burnt matches. Sulphur dioxide is one of the main components of acid rain. The main source of sulphur dioxide is probably coal-fired power generation, since this often uses low grade coal, which may have a high level of sulphur in it.

Sulphur dioxide is also produced during the combustion of hydrogen sulphide, which may be present in biogas from animal sources. Sulphur dioxide levels are regulated in most countries due to its aggressive nature and effect on plant and animal life. Many of the sources of sulphur dioxide are not easily avoided, so the main remedy is filtration to remove sulphur dioxide and other toxic components.

FEATURES

SO_2 is a colourless gas. It smells like burnt matches. It can be oxidized to sulphur trioxide, which in the presence of water vapour is readily transformed to sulphuric acid mist. SO_2 can be oxidized to form acid aerosols. SO_2 is a precursor to sulphates, which are one of the main components of respirable particles in the atmosphere.

SOURCES

Approximately 69 per cent of the SO_2 emitted in Ontario in 2006 came from smelters and utilities. Downstream petroleum industry and other industrial processes each accounted for approximately 9 per cent of SO_2 emissions.

Lesser sources of SO_2 include transportation, cement and concrete, and miscellaneous.

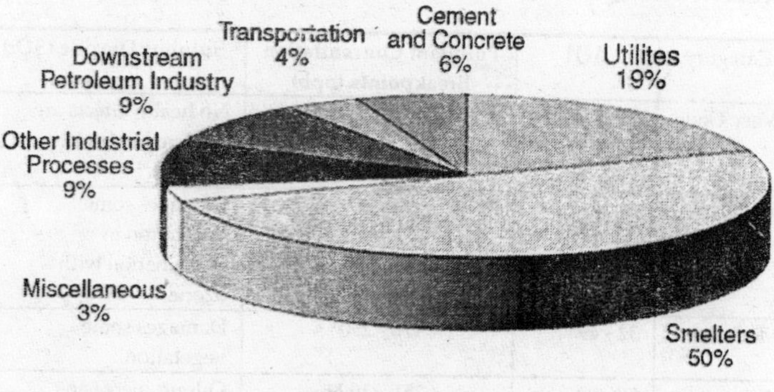

Fig. Pie Chart

Table. Ontario Sulphur Dioxide Emissions by Sector

Category	Per cent
Smelters	50%
Utilities	19%
Downstream Petroleum Industry	9%
Other Processes	9%
Cement and Concrete	6%
Transportation	4%
Miscellaneous	3%

EFFECTS OF SO_2

Health effects caused by exposure to high levels of SO_2 include breathing problems, respiratory illness, changes in the lung's defences, and worsening respiratory and cardiovascular disease. People with asthma or chronic lung or heart disease are the most sensitive to SO_2. It also damages trees and crops. SO_2, along with nitrogen oxides, are the main precursors of acid rain.

This contributes to the acidification of lakes and streams, accelerated corrosion of buildings and reduced visibility. SO_2 also causes formation of microscopic acid aerosols, which have serious health implications as well as contributing to climate change. The following table shows the health effects of different Air Quality Index levels caused by sulphur dioxide.

Category	AQI	Pollutant Concentration Breakpoints (ppb)	Sulphur Dioxide (SO_2)
Very Good	0 - 15	0 – 79	No health effects are expected in healthy people.
Good	16 - 31	80 – 169	Damages some vegetation in combination with ozone.
Moderate	32 - 49	170 - 250*	Damages some vegetation.
Poor	50 - 99	251 – 1999	Odour; increasing vegetation damage.
Very Poor	100 or over	2000 or over	Increasing sensitivity for asthmatics and people with bronchitis.

BASIC CHEMICAL AND PHYSICAL DATA

Empirical formula	SO_2
Rel. molecular mass	64.06 g
Density	1.46 g/cm^3 at -10°C (liquid); 2.93 g/l at 20°C (gas)
Relative gas density	2.26
Boiling point	–10°C
Melting point	–75.5°C
Vapour pressure	331 kPa at 20°C, 462 kPa at 30°C, 842 kPa at 50°C
Odour threshold	0.3 – 1 ppm (in air)
Solvolysis/solubility	in water: 112.7 g/l at 20°C (1013 mbar); 228.3 g/l at 0°C (1013 mbar); readily soluble in alcohol, benzene, acetone, carbon tetrachloride; fully miscible with ether, carbon disulphide, chloroform, glycol
Conversion factors	1 ppm = 0.376 mg/m^3 1 mg/m^3 = 2.663 ppm

ORIGIN AND USE

Usage

There are many uses of sulphur dioxide. It is used e.g. as a reducing agent in metallurgy, as a coolant in the refrigeration industry, as a disinfectant and bleach, in the preservation of foodstuffs, for dechlorination and as a fumigant.

Sulphur dioxide is one of the most important compounds in the chemical industry. 98% of technical SO_2 is used in the production of sulphur trioxide as a precursor of sulphuric acid.

Origin/Derivation

Sulphur dioxide is released naturally into the atmosphere from volcanoes and combustion processes. The anthropogenic impact on the environment primarily results from the combustion of sulphurous fossil fuels (e.g. coal, oil, natural gas) in power and heating plants, in industry, in household use and in traffic.

The technical product is made from elemental sulphur, pyrite, sulphide ores of non-ferrous metals, gypsum, anhydrite and flue gases.

Production Figures

Excluded production from elemental sulphur and pyrites in 1,000 t of sulphur:

WorldwideM: 5,820	*Soviet Union*: 1,700
United States: 1,380	*Japan*: 1,370

- Production from pyrites in 1,000 t of sulphur:
  Worldwide: 11,000
- Production from metal ores and sulphur in 1,000 t of sulphur:
  Worldwide: 20,000

Emission Figures (Estimated)

Total emissions in Germany were calculated at 2.3×10^6

tons approximately. Natural emission have been estimated at 750×10^6 t worldwide, whereas anthropogenic emissions amounted to about 100×10^6 t.

Toxicity

Humans:	25 μg/m³ (annual average)	increased frequency of diseases of lower respiratory tract (acc. UN-ECE, 1984)
	225 μg/m³ (annual average)	increased frequency of respiratory symptoms; reduced pulmonary function in
		5 year olds (acc. UN-ECE, 1984)
	200 μg/m³ (daily max., 30 min)	significant increase in pseudocroup in children (acc. AFRL, 1987)
	200 μg/m³ (24 h values)	increased mortality amongst elderly people (acc. AFRL, 1987)
	1.3 mg/m³ (40 min)	constriction of the respiratory tract amongst people suffering from asthma (acc. AFRL, 1987)
	53.3 mg/m³ (10-30 min)	severe, extremely unpleasant irritation symptoms (acc. DFG, 1988)
	133.2 mg/m³ (60 min)	severe irritation of mucous membranes, pulmonary haemorrhage and oedema,
		laryngospasm with danger of asphyxiation (acc. DFG, 1988)

Mammals

Mouse:	LC_{50} 346 mg/m³ (24 h)
Mouse:	LC 1,598 mg/m³ (5 h)
Mouse:	LC 2,130 mg/m³ (20 min)
Rabbit:	LC_{50} 679 mg/m³ (24 h)
Rabbit:	LC (after 7 d) 2130 mg/m³ (1 h)
Hamster:	LC 1,065 mg/m³ (6 h)
Guinea pig:	LC_{50} 1,076 mg/m³ (24 h)
Insects:	LC 2 vol% (6 h)

Various species	>20 μg/m³ (annual average, visible damage)
Fir	30-40 μg/m³ (annual average, damage)
Fir	50-70 μg/m³ (annual average, severe damage)
Cultivated plants	50 μg/m³ (90 d, damage)
Pines (Ruhr area)	>80 μg/m³ (average, vegetation period, initial damage)
Various species	2.7-5.5 mg/m³ (a few hours, acute damage)[1)]

Sensitivity of Higher Plants

very sensitive	Bean	Blackcurrant	Sweet pea	Walnut
	Douglas fir	Clover	Spinach	
	Pea	Lupin	Gooseberry	
	Fir	Lucerne	Pine	
Sensitive	Linden	Pine	Oats	Bean
	Copper beech	Weymouth pine	Rye	Rape
	Apple	Larch	Wheat	
	Hazelnut	Barley	Lettuce	
less sensitive	Maple	Potato	Plane	Tomato
	Beech	Cabbage	Plum family	Juniper
	Yew	Leek	Rhododendron	Willow
	Oak	Arborvitae	Robinia	Vine
	Strawberry	Maize	Turnip	
	Aldo	Carrot	False cypress	
	Lilac	Poplar	Black pine	

CHARACTERISTIC EFFECTS

Humans/Mammals

Keratitis, breathing difficulties, inflammation of respiratory organs and irritation of eyes due to the formation of sulphurous acid on the moist mucous membranes. Disturbances of consciousness, pulmonary oedema, bronchitis, heart failure and circulatory collapse. Similar effects with sulphur trioxide (SO_3)

Plants

Visible damage to parts of plants above ground level due to direct action: SO_2 enters the leaves via the stomata. It physiologically and biochemically impairs the photosynthesis, the respiration and the transpiration due to its detrimental effect on the pore aperture mechanism. Indirect damage is above all due to soil acidification (damage to mycorrhiza) and results in stunted growth.

ENVIRONMENTAL BEHAVIOUR

Water: Sulphur dioxide ingresses into surface water and groundwater through dry and wet deposition. The aqueous solution reacts as a strong acid. In Germany, SO_2 is classed as hazardous to water as are sulphuric acid and sulphurous acid.

Air: Sulphur dioxide binds moisture from the air and forms aerosols of sulphuric and sulphurous acid which are deposited as acid rain.

The aerosol formation and its dwell time in air depend on the meteorological conditions and on the presence of catalytic impurities in the air. The average dwell time in the atmosphere is approx. 3 – 5 days. Thus, sulphur dioxide may also be transported over long distances.

Soil: Dry and wet depositions from the atmosphere are the major sources of sulphate accumulation in soil. Dry deposition particles chiefly consist of $(NH_4)_2SO_4$, $(NH_4)_3H(SO_4)_2$, $CaSO_4$, $MgSO_4$ with a small percentage of organic sulphur compounds.

SO_2 and its transformation products are the major sources of soil acidification, particularly if the buffer system of the soil is incapable of neutralising the acid which is either directly deposited or produced by the conversion of solid sulphates. The damage is not substance-specific. Almost all reactions in soil depend on the pH: Both the desorption of numerous substances with adverse effects as well as the leaching of nutrients increase with the acidification of the soil.

Degradation, decomposition products, half-life: As described above, sulphur dioxide is readily oxidised and very reactive. Sulphuric and sulphurous acid are the most important reaction products relevant to the environment.

Synergisms/antagonisms: Numerous experiments have been performed in this field generally under standardised conditions. It is however not possible to provide quantitative information relating to natural circumstances on account of the complexity of the factors involved and the courses of action concerned. Nonetheless, it is certain that the effect of SO_2 is more than additively enhanced in combination with other pollutant gases such as NO_x or HF.

Assessment/Comments

The laboratory animals used to date for toxicity experiments are obviously far less sensitive to sulphur dioxide than humans. The most sensitive animal species (guinea pig) withstands - even over long periods - concentrations which are already intolerable to humans in the short term.

Sulphur dioxide is one of the chemicals for which there is a wealth of legislation. Limit and approximate values with differing points of reference are available from numerous countries. In comparison to the values for numerous other substances, the figures for SO_2 are subject to relatively rapid change.

When comparing the extensive range of values available, it is important to take account of the method of calculation (median, arithmetic mean, time frame, percentile etc.). The listed values for the Netherlands and the EC are good examples.

UBA compares technical installations, provides information on the sulphur content of raw materials from various countries and outlines different scenarios.

We have to distinguish between SO_2 produced for industrial processes (e.g. production of sulphuric acid) and SO_2 which is emitted. Although most of the SO_2 is of natural origin, care should be taken to reduce the emissions caused by humans, especially those from combustion processes.

SULPHURIC ACID

Sulfuric acid is an oily, colorless liquid with no odor. It is a key component in batteries, wastewater treatment, ore production, and fertilizer creation. Sixty-five per cent of all fertilizers available on the market today are created from a combination of sulfuric acid and other chemicals. It is created from inorganic materials through a process that utilizes a special type of chemical reaction. More than 40 million tons of sulfuric acid are produced annually in the United States alone.

Sulfuric acid was discovered in the 8th century by an alchemist named Jabir Ibn Hayyan. When sulfur compounds are burned, sulfur dioxide gas is left behind. When sulfur dioxide gas is heated to 450 degrees, the combination turns into sulfur trioxide. Jabir Ibn Hayyan mixed sulfur trioxide with water to create sulfuric acid. A century later, Ibn Zakariya al-Razi, a Persian physician and alchemist, continued studying the acid and discovered it had the ability to destroy iron and copper oxide.

A variety of substances can be mixed with sulfuric acid in order to create new products. Sulfur can be mixed with forms of aluminum to create aluminum sulfates. When aluminum sulfates are mixed with wood pulp, the fibers bond together and create paper. Aluminum sulfates mixed with dioxides create a mixture that takes impurities out of water. These aluminum dioxides are commonly used at water treatment plants.

Cyclohexanoneoxime and sulfuric acid combine to make caprolactum. Caprolactum is used to make nylon. Hydrochloric acid is another creation derived from sulfuric acid that is commonly used in the world today. In addition, sodiums and sulfuric acids are combined to help refine petroleum in order to create gasoline.

Due to the acid's unstable properties, it is essential for those combining sulfuric acid with water to add it to the water and not the other way around. If water is poured into sulfuric acid, a volatile explosion of boiling water can be created. Though sulfuric acid is non-flammable, it can create hydrogen gas, which is highly volatile.

Chapter 3

Occurrence and Properties in Plants

SODIUM

CHARACTERISTICS

Sodium is a soft, silvery-white metal. It is soft enough to cut with the edge of a coin. Freshly cut surfaces oxidize rapidly in air to form a dull, oxide coating. Sodium burns in air with a brilliant yellow flame.

Sodium floats on water, because its density is lower than water's. It also reacts vigorously with water - violently if more than a small amount of sodium meets water - to produce sodium hydroxide and hydrogen gas. Sodium reacts with water more vigorously than lithium and less vigorously than potassium. Explosions occur when the heat generated by the sodium-water reaction ignites the resulting hydrogen gas.

USES

Metallic sodium is vital in the manufacture of esters and in the preparation of organic compounds. The metal may be used to improve the structure of certain alloys, descale metal, and purify molten metals. An alloy of sodium with potassium, NaK, is an important heat transfer agent.

Metallic sodium is used in the manufacture of sodamide and esters, and in the preparation of organic compounds. The metal also may be used to modify alloys such as aluminum-silicon by improving their mechanical properties and fluidity.

Sodium is used to descale (smooth the surface of) metals and to purify molten metals.

Sodium vapour lamps are highly efficient in producing light from electricity and are often used for street lighting in cities. Sodium is used as a heat transfer agent; for example, liquid sodium is used to cool nuclear reactors. Sodium chloride (table salt, NaCl) is vital for good nutrition. Sodium ions facilitate transmission of electrical signals in the nervous system and regulate the water balance between body cells and body fluids.

Physical Prperties

- *Atomic Number*:11
- *Atomic Radius*:144.4 pm
- *Atomic Symbol*:Na
- *Melting Point*: 97.8°C
- *Atomic Weight*:22.98977
- *Boiling Point*:883°C
- *Electron Configuration*:[Ne]3s1
- *Oxidation States*:1
- *History*: Long recognized in compounds, sodium was first isolated by Davy in 1807 by electrolysis of caustic soda.
- *Sources*: Sodium is present in fair abundance in the sun and stars. The D lines of sodium are among the most prominent in the solar spectrum. Sodium is the fourth most abundant element on earth, comprising about 2.6% of the earth's crust; it is the most abundant of the alkali group of metals. It is now obtained commercially by the electrolysis of absolutely dry fused sodium chloride. This method is much cheaper than that of electrolyzing sodium hydroxide, as was used several years ago.
- *Compounds*: The most common compound is sodium chloride (table salt), but it occurs in many other minerals, such as soda niter, cryolite, amphibole, zeolite, etc.
- *Properties*: Sodium, like every reactive element, is

never found free in nature. Sodium is a soft, bright, silvery metal which floats on water. Decomposition in water results in the evolution of hydrogen and the formation of the hydroxide. It may or may not ignite spontaneously on water, depending on the amount of oxide and metal exposed to the water. It normally does not ignite in air at temperatures below 115°C.

- *Compounds*: Sodium compounds are important to the paper, glass, soap, textile, petroleum, chemical, and metal industries.

  Soap is generally a sodium salt of certain fatty acids. The importance of common salt to animal nutrition has been recognized since prehistoric times. Among the many compounds that are of the greatest industrial importance are common salt (NaCl), soda ash (Na_2CO_3), baking soda ($NaHCO_3$), caustic soda (NaOH), Chile saltpeter ($NaNO_3$), di- and tri-sodium phosphates, sodium thiosulfate (hypo, $Na_2S_2O_3 . 5H_2O$), and borax ($Na_2B_4O_7 . 10H_2O$).
- *Isotopes*: Thirteen isotopes of sodium are recognized.
- *Handling*: Sodium metal should be handled with great care. It cannot be maintained in an inert atmosphere and contact with water and other substances with which sodium reacts should be avoided.
- *Title Picture*: Sodium reflects yellow light, as observed in astronomical phenomena.

SODIUM CHLORIDE

OCCURRENCE

The classic case of ionic bonding, the sodium chloride molecule forms by the ionization of sodium and chlorine atoms and the attraction of the resulting ions.

An atom of sodium has one 3s electron outside a closed shell, and it takes only 5.14 electron volts of energy to remove that electron. The chlorine lacks one electron to fill a shell, and releases 3.62 eV when it acquires that electron (it'selectron affinity is 3.62 eV).

This means that it takes only 1.52 eV of energy to donate one of the sodium electrons to chlorine when they are far apart. When the resultant ions are brought closer together, their electric potential energy becomes more and more negative, reaching -1.52 eV at about 0.94 nm separation.

This means that if neutral sodium and chlorine atoms found themselves closer than 0.94 nm, it would be energetically favorable to transfer an electron from Na to Cl and form the ionic bond.

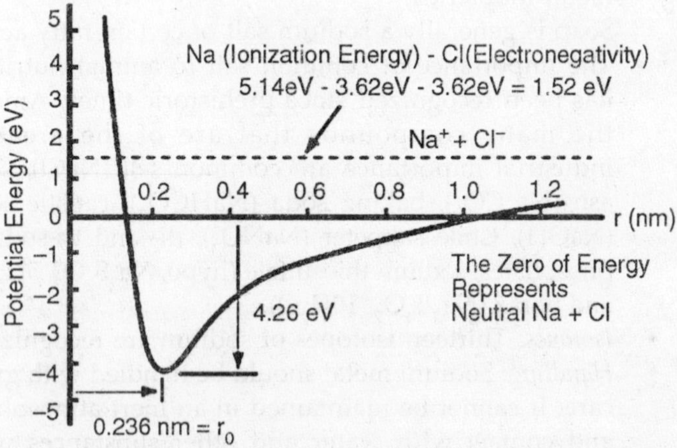

The potential energy curve shows that there is a minimum at 0.236 nm separation and then a steep rise in potential which represents a repulsive force. This repulsive force is more than just an electrostatic repulsion between the electron clouds of the two atoms. It has a quantum mechanical character rooted in the Pauli exclusion principle, and is often called just the"exclusion principle repulsion".

When the ions are widely separated, the wavefunctions of their core electrons do not significantly overlap and they can have identical quantum numbers. As they get closer, the increasing overlap of the wavefunctions causes some to be forced into higher energy states.

No two electrons can occupy the same state, so as a new set of energy states is formed for the composite, two-nucleus system, the lower energy states are filled and some of the electrons are pushed into higher states. This requires energy and is experienced as a repulsion, preventing the ions from coming any closer to each other.

The potential diagram above is for gaseous NaCl, and the environment is different in the normal solid state where sodium chloride (common table salt) forms cubical crystals. The ion separation is 0.28 nm, somewhat larger than that in the gaseous state.

A major part of the study of molecular structure is the description of the chemical bonds which are formed between atoms. The classic studies are the extremes of ionic bonding in sodium chloride and covalent bonding in the hydrogen molecule.

SODIUM BICARBONATE

OVERVIEW

Sodium Bicarbonate (SOE dee um; bye KAR bon ate) is an antacid. It is used to treat acid indigestion and heartburn caused by too much acid in the stomach.

This medicine may be used for other purposes; ask your health care provider orpharmacist if you have questions.

They need to know if you have any of these conditions:

- Cushing's syndrome
- Heart problems
- Kidney disease
- Low levels of calcium or potassium in the blood
- Low salt or sodium diet
- Stomach ulcer
- An unusual or allergic reaction to sodium bicarbonate, other medicines, foods, dyes, or preservatives
- Pregnant or trying to get pregnant
- Breast-feeding

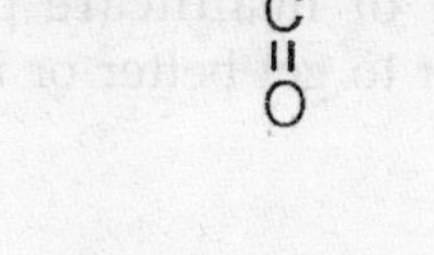

USING THIS MEDICINE

Take this medicine by mouth. Dissolve it in a glass of water before taking. Do not take tablets whole. Follow the directions on the package label.

Take your medicine at regular intervals. Do not take it more often than directed. Talk to your pediatrician regarding the use of this medicine in children. Special care may be needed. Patients over 60 years old may have a stronger reaction and need a smaller dose. *Over dosage*: If you think you have taken too much of this medicine contact a poison control centre or emergency room at once. This medicine is only for you. Do not share this medicine with others.

Do not take this medicine with any of the following medications:

- *Ammonium chloride*: This medicine may also interact with the following medications:
- Antibiotics like ciprofloxacin, doxycycline, Tetracycline
- Aspirin and aspirin-like medicines
- Delavirdine
- Ephedra, Ma Huang
- Itraconazole, ketoconazole
- Lithium
- Methenamine
- Norfloxacin
- Quinidine
- Quinine
- Stimulants like amphetamine, dexmethylphenidate, and others
- Tolmetin

This list may not describe all possible interactions. Give your health care provider a list of all the medicines, herbs, non-prescription drugs, or dietary supplements you use. Also tell them if you smoke, drink alcohol, or use illegal drugs. Some items may interact with your medicine.

Tell your doctor or healthcare professional if your symptoms do not start to get better or if they get worse. Do not take this medicine for more than 2 weeks unless your doctor directs you to.

Side Effect

Side effects that you should report to your doctor or health care professional as soon as possible:

- Allergic reactions like skin rash, itching or hives, swelling of the face, lips, or tongue
- Confusion
- Dizziness
- Muscle pain
- Nausea, vomiting
- Seizures
- Swelling of the feet and legs
- Unusually weak or tired

Side effects that usually do not require medical attention (report to your doctor or health care professional if they continue or are bothersome):

- Bloating and stomach gas
- Increased thirst
- Stomach cramps

CALCIUM SULPHATE

OCCURRENCE

Calcium sulphate Chemical compound ($CaSO_4$) that occurs naturally as the mineral anhydrite. The hardness of water results when calcium sulphate dissolves in water. The hydrated form is gypsum, which loses water when heated to form plaster of Paris.

LIGHT CALCIUM SULPHATE

This product is an amorphous calcium sulphate and unlike the natural and its processed forms (gypsum etc.), it is very fluffy, white, pure and fine. Hence the product is highly suitable for its calcium contents utilization and other properties in applications where chemical purity and physical properties are important.

APPLICATIONS

Food Items	Portland
Pharmaceuticals	Cosmetics
Dyes & Pigments	Ceramics
Paints, Paper, Rubber as a filler	Chemicals
Plastics	Fertilizers
Beer	Desiccant

Table. Typical Physiochemical Properties Of Calcium Sulphate

Property	Unit	Value
Surface Area (BET)	m^2/gm	
Bulk Density	gm/cc	0.5±0.05
Water Absorption	%	55–65
pH (2% acq. Solution)	–	8.0±0.5
Sp. Gravity	–	2.32
Moisture (at 105°C)	%max.	1.0
Loss on Ignition (900°C for 2 hrs)	%max.	4
Residue on 300 mesh	%max.	0.5

CALCIUM NITRATE

OCCURRENCE

Calcium nitrate is a common laboratory reagent, used instead of calcium chloride as a source of calcium ions in reactions and syntheses where the chloride ion would interfere. Danger! Strong oxidizer! Contact with other material may cause fire. Avoid contact with eyes, skin, and clothing.

Calcium Nitrate, Monohydrate. $Ca(NO_3)_2 \bullet H_2O$; mol. wt. 182.11. Anhydr. salt 90.11%, H_2O 9.89%, Ca 22.01%, N 15.38%, NO_3 68.1%, CaO 30.79%, N_2O_5 59.31%.

Deliquesc. granules. m. about 560°. Very soluble in water, heat being evolved; freely soluble in alcohol; almost insoluble in concentr. HNO_3.

This is sometimes a component of mineral waters. It should be preserved in well-stoppered bottles, as it readily absorbs moisture from the air It is best kept in solution, or, still better, prepared for immediate use from calcium carbonate and nitric acid.

Wherever possible it is produced by mutual decomposition of chloride of calcium and nitrate alkalies, within the fountain; but when its separate addition is necessary it is best prepared as follows:

PREPARATION

Take 61 parts by weight of calcium carbonate, marble dust, etc., add 658.5 parts by weight of nitrjc acid diluted to 10 ° Baume of 1.069 specific gravity at 15° (J., stir well and add distilled water to make it 1000 parts. Filter and keep in stoppered bottle. This repra sents then a 10 per cent, solution of calcium nitrate, which is kept in stock ready for use.

Proportion 10 to 1. One part by weight of commercial calcium nitrate dissolved in nine parts of distilled water (10 per cent.). Filter. Proportion 10 to 1.

IRON

OCCURRENCE

Much of the iron in the body is contained in hemoglobin. Hemoglobin is the component of red blood cells that enables them to carry oxygen and deliver it to the body's tissues. Iron also is an important component of muscle cells. It is also necessary for the formation of many enzymes in the body.

The body recycles iron: When red blood cells die, the iron in them is returned to the bone marrow to be used again in new red blood cells. A small amount of iron is lost each day, mainly in cells shed from the lining of the intestine. This amount is usually replaced by the 1 to 2 milligrams of iron absorbed from food each day.

Food contains two types of iron:

- *Heme iron*: Animal products contain heme iron. It is absorbed much better than nonheme iron.

- *Nonheme iron*: Most foods and iron supplements contain nonheme iron. It accounts for more than 85% of iron in the average diet. However, less than 20% of nonheme iron that is consumed is absorbed into the body. Nonheme iron is absorbed better when it is consumed with animal protein and with vitamin C.

DEFICIENCY

- Iron deficiency usually results from loss of blood in adults but, in children and pregnant women, may result from an inadequate diet.
- Anemia develops, making people appear pale and feel weak and tired.
- Doctors base the diagnosis on symptoms and blood test results.
- Doctors look for a source of bleeding, and if one is identified, they treat it.
- Iron supplements, usually taken by mouth, are often needed.

Iron deficiency is one of the most common mineral deficiencies in the world. It causes anemia in men, women, and children. In adults, iron deficiency is most commonly caused by loss of blood.

In premenopausal women, monthly menstrual bleeding may cause the deficiency. In men and postmenopausal women, iron deficiency usually indicates bleeding, most often in the digestive tract—for example, from a bleeding ulcer or a polyp in the colon. Chronic bleeding due to colon cancer is a serious cause in middle-aged and older people.

Iron deficiency may result from an inadequate diet, primarily in infants and small children, who need more iron because they are growing.

Adolescent girls who do not eat meat are at risk of developing iron deficiency because they are growing and starting to menstruate. Pregnant women are at risk of this deficiency because the growing fetus requires large amounts of iron.

SYMPTOMS

When iron reserves in the body are exhausted, anemia develops (Anemia: Iron Deficiency Anemia). Anemia causes paleness, weakness, and fatigue. People usually do not notice how pale they are because it happens so gradually. Concentration and learning ability may be impaired.

When severe, anemia may cause shortness of breath, dizziness, and a rapid heart rate. Occasionally, severe anemia causes chest pain and heart failure. Menstrual periods may stop. In addition to anemia, iron deficiency may cause pica (a craving for nonfoods such as ice, dirt, or pure starch), spoon nails (thin, concave fingernails), and leg cramps at night. Rarely, iron deficiency causes a thin membrane to grow across part of the esophagus, making swallowing difficult.

DIAGNOSIS

Iron deficiency is diagnosed based on symptoms and blood test results. Results include a low level of hemoglobin (which contains iron), a low hematocrit (the percentage of blood volume that is red blood cells), a low number of red blood cells, and the presence of abnormally small red blood cells. Blood tests also include the following:

- *Transferrin*: Transferrin is the protein that carries iron in blood when iron is not inside red blood cells. If the percentage of iron in transferrin is less than 10%, iron deficiency is likely.
- *Ferritin*: Ferritin is a protein that stores iron. Iron deficiency is confirmed if the ferritin level is low.

However, the ferritin level may be normal or high when iron deficiency is present if people have inflammation, an infection, cancer, or liver damage.

Occasionally, a bone marrow examination is needed to make the diagnosis. A sample of bone marrow cells is removed, usually from the hipbone, through a needle and examined under a microscope to determine the iron content.

TREATMENT

Because the most common cause of iron deficiency in

adults is excessive bleeding, doctors first look for a source of bleeding. If the source is excessive menstrual bleeding, drugs, such as oral contraceptives (birth control pills), may be needed to control it. Surgery may be needed to repair a bleeding ulcer or remove a polyp in the colon. A blood transfusion may be necessary if the anemia is severe.

Normal dietary intake of iron may not be sufficient to replace lost iron (because less than 20% of iron in a typical diet is absorbed into the body). Thus, most people with iron deficiency need to take iron supplements by mouth usually once or twice a day. Iron in supplements is absorbed best when taken on an empty stomach, 30 minutes before meals or 2 hours after meals, particularly if the meals include foods that reduce the absorption of iron (such as vegetable fibers, phytates, bran, coffee, and tea). However, taking iron supplements on an empty stomach can cause indigestion and constipation.

So some people must take the supplements with meals. Antacids and calcium supplements can also reduce iron absorption. Consuming vitamin C in juices or taking it as a supplement enhances iron absorption, as does eating small amounts of meat, which contains the more easily absorbed form of iron (heme iron). Iron supplements almost always turn stools black—a harmless side effect.

Rarely, iron is given by injection. Injections are necessary for people who cannot tolerate tablets or for a few people who cannot absorb enough iron from the digestive tract.

Correcting iron deficiency anemia usually takes 3 to 6 weeks, even after the bleeding has stopped. After the anemia is corrected, an iron supplement should be taken for 6 months to replenish the body's reserves. Blood tests are usually done periodically to determine whether people are receiving enough iron and to check for continued bleeding.

Women who are not menstruating and men should not take iron supplements or multiple vitamins with iron unless they are specifically instructed to do so by a doctor. Taking such supplements can make diagnosing bleeding from the intestine difficult. Such bleeding may be due to serious disorders, including colon cancer.

Because a developing fetus requires iron, iron supplements are recommended for most pregnant women. Most babies, particularly those who are premature or who have a low birth weight, need an iron supplement. It is given as an iron-fortified formula or, to breastfed babies, as a separate liquid supplement.

EXCESS

Excess iron can accumulate in the body. Causes include the following:

- Repeated blood transfusions
- Iron therapy given in excessive amounts or for too long
- Chronic alcoholism
- An overdose of iron
- A hereditary disorder called hemochromatosis

Excess iron consumed all at once causes vomiting, diarrhea, and damage to the intestine and other organs. Excess iron consumed over a period of time may damage coronary arteries. Often, deferoxamine is given intravenously. This drug binds with iron and carries it out of the body in urine. Hemachromatosis is treated with bloodletting (phlebotomy).

HEMOCHROMATOSIS

In hemochromatosis, a hereditary disorder, too much iron is absorbed, resulting in the accumulation of iron in the body.

In the United States, over 1 million people have hemochromatosis. The disorder is potentially fatal but usually treatable. The gene associated with hemochromatosis has been identified.

- Iron can accumulate and damage any part of the body.
- People may develop symptoms of cirrhosis or diabetes or simply feel tired.
- Blood tests identify people who require genetic testing, which can confirm the disorder.
- Bloodletting, done periodically, can prevent further damage.

Symptoms

Usually, symptoms develop gradually, often not appearing until middle age or later. In women, symptoms usually start after menopause because the loss of iron during menstrual bleeding and the increased requirement for iron during pregnancy compensate to some degree.

Symptoms vary because iron accumulation can damage any part of the body, including the brain, liver, pancreas, lungs, or heart. The first symptoms, particularly in men, may be those of cirrhosis (due to liver damage) or those of diabetes (due to pancreas damage).

Or, the first symptoms, particularly in women, may be vague and affect the whole body. Fatigue is an example. Liver dysfunction is the most common problem. The following problems can also occur:

- Bronze-colored skin
- Heart failure (occasionally)
- Joint pains
- Increased risk of liver cancer
- Infertility
- An underactive thyroid gland (hypothyroidism)
- Chronic fatigue

In many men, levels of male hormones decrease. Erectile dysfunction (impotence) may occur. Hemochromatosis can worsen neurologic disorders that are already present.

DIAGNOSIS

Identifying hemochromatosis based on symptoms may be difficult. However, blood tests to measure the levels of iron and two other substances can identify people who should be further evaluated. These substances are ferritin, a protein that stores iron, and the iron in transferrin, the protein that carries iron in blood when iron is not inside red blood cells.

If the ferritin level and percentage of iron in transferrin are high, genetic testing is usually done to confirm the diagnosis. A liver biopsy may be necessary to determine whether the liver has been damaged. Genetic testing is recommended for people with hemochromatosis and all of their first-degree relatives (siblings, parents, and children).

TREATMENT

Usually, bloodletting (phlebotomy) is the best treatment. It prevents additional organ damage but does not reverse existing damage.

Bloodletting is done once or sometimes twice a week. Each time, about 500 milliliters (1 pint) of blood is removed until the iron level and percentage of iron in transferrin are normal. Bloodletting is then done periodically to keep these substances at normal levels.

ALUMINIUM SULPHATE

OCCURRENCE

Aluminium sulfate or Aluminium Sulphate, written as $Al_2(SO_4)_3$ or $Al_2O_{12}S_3$ Aluminium sulfate is an industrial chemical used as a flocculating agent in the purification of drinking water and waste water treatment plants, and also in paper manufacturing.

Aluminium sulfate is sometimes incorrectly referred to as alum but alums are closely related compounds typified by $KAl(SO_4)_2 \cdot 12H_2O$. The anhydrous form occurs naturally as a rare mineralmillosevichite, found e.g. in volcanic environments and on burning coal-mining waste dumps.

Aluminium sulfate is rarely, if ever, encountered as the anhydrous salt. It forms a number of different hydrates, of which the hexadecahydrate $Al_2(SO_4)_3 \cdot 16H_2O$ and octadecahydrate $Al_2(SO_4)_3 \cdot 18H_2O$ are the most common. The heptadecahydrate, whose formula can be written as $[Al(H_2O)_6]_2(SO_4)_3 \cdot 5H_2O$, occurs naturally as the mineral alunogen.

PREPARATION

Aluminium sulfate may be made by aluminium hydroxide, $Al(OH)_3$, in sulfuric acid, H_2SO_4:

$$2\ Al(OH)_3 + 3\ H_2SO_4 \rightarrow Al_2(SO_4)_3 \cdot 6H_2O$$

USES

Aluminium Sulfate is used in water purification and as a mordant in dyeing and printing textiles. In water purification, it causes impurities to coagulate which are removed as the particulate settles to the bottom of the container or more easily filtered.

This process is called coagulation or flocculation. When dissolved in a large amount of neutral or slightly-alkaline water, aluminium sulfate produces a gelatinous precipitate of aluminium hydroxide, $Al(OH)_3$. In dyeing and printing cloth, the gelatinous precipitate helps the dye adhere to the clothing fibers by rendering the pigment insoluble.

Aluminium sulfate is sometimes used to reduce the pH of garden soil, as it hydrolyzes to form the aluminium hydroxide precipitate and a dilute sulfuric acid solution. An example of what changing the pH level of soil can do to plants is visible when looking at the Hydrangea macrophylla.

The gardener can add Aluminum sulfate to the soil to reduce the pH level which in turn will result in the flowers of the Hydrangea turning a different colour. Aluminium sulfate is the active ingredient of some antiperspirants; however, beginning in 2005 the US Food and Drug Administration no longer recognized it as a wetness reducer.

Aluminium sulfate is usually found in baking powder, where there is controversy over its use due to concern regarding the safety of adding aluminum to the diet. In construction industry it is used as waterproofing agent and accelerator in concrete.

Another use is a foaming agent in fire fighting foam. It is also used in styptic pencils, and pain relief from stings and bites; it is the active ingredient in popular pain relief products such as Stingose.

CHEMICAL REACTIONS

It combines with water forming hydrated salts of various compositions. Produces calcium aluminate with the evolution of sulfate when calcined with calcium sulfate:

$$Al_2(SO_4)_3 + CaSO_4 \rightarrow Ca(AlO_2)_2 + 4\ SO_3$$

Aluminium sulfate reacts with sodium bicarbonate to which foam stabilizer has been added, typically made from saponin, licorice root, or hydrolyzed animal protein in aqueous solution, forming fire-extinguishing foams, producing carbon dioxide:

$$Al_2(SO_4)_3 + 6\ NaHCO_3 \rightarrow 2Al(OH)_3 + 3\ Na_2SO_4 + 6\ CO_2$$

The carbon dioxide is trapped by the foam stabilizer and creates a thick foam which will float on top of hydrocarbon fuels and seal off access to atmospheric oxygen, smothering the fire. Chemical foam was unsuitable for use on polar solvents such as alcohol, as the fuel would mix with and break down the foam blanket.

The carbon dioxide generated also served to propel the foam out of the container, be it a portable fire extinguisher or fixed installation using hose lines. Chemical foam is considered obsolete in the country and has been replaced by synthetic mechanical foams, such as AFFF which have a longer shelf life, are more effective, and more versatile, although some counties such as Japan and India continue to use it.

NITROGEN CYCLE

OVERVIEW

Nitrogen (N) is an essential component of DNA, RNA, and proteins, the building blocks of life. All organisms require nitrogen to live and grow.

Although the majority of the air we breathe is N_2, most of the nitrogen in the atmosphere is unavailable for use by organisms. This is because the strong triple bond between the N atoms in N_2 molecules makes it relatively inert.

In fact, in order for plants and animals to be able to use nitrogen, N_2 gas must first be converted to more a chemically available form such as ammonium (NH_4^+), nitrate (NO_3^-), or organic nitrogen (e.g. urea - $(NH_3)_2CO$). The inert nature of N_2 means that biologically available nitrogen is often in short supply in naturalecosystems, limiting plant growth and biomass accumulation. Nitrogen is an incredibly versatile element, existing in both inorganic and organic forms as well as many different oxidation states. The movement of nitrogen between the atmosphere, biosphere, and geosphere in different forms is described by the nitrogen cycle, one of the major biogeochemical cycles.

Similar to the carbon cycle, the nitrogen cycle consists of various storage pools of nitrogen and processes by which the pools exchange nitrogen (arrows).

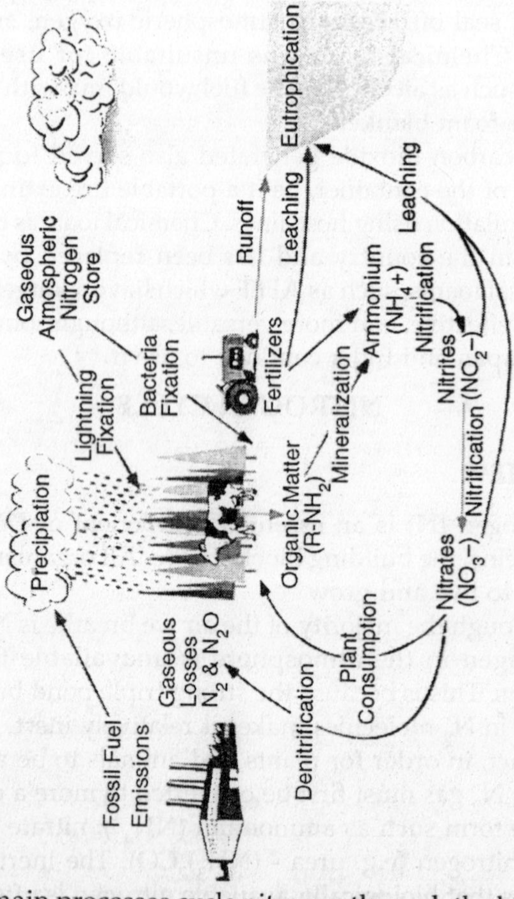

Five main processes cycle nitrogen through the biosphere, atmosphere, and geosphere: nitrogen fixation, nitrogen uptake (organismal growth), nitrogenmineralization (decay), nitrification, and denitrification. Microorganisms, particularly bacteria, play major roles in all of the principal nitrogen transformations.

As microbially mediated processes, these nitrogen transformations tend to occur faster than geological processes like plate motion, a very slow, purely physical process that is a part of the carbon cycle.

Instead, rates are affected by environmental factors that influence microbial activity, such as temperature, moisture, and resource availability.

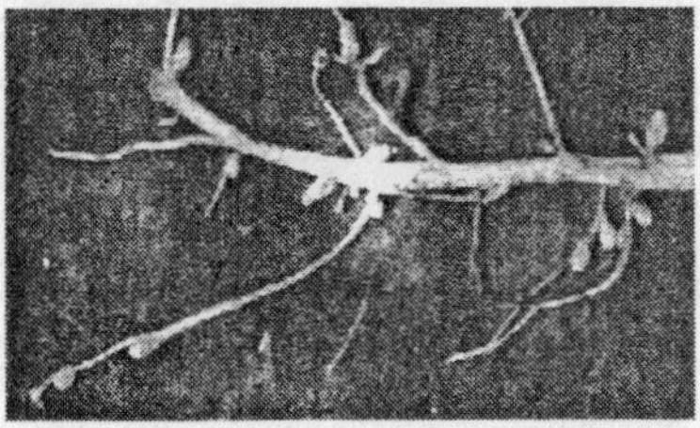

NITROGEN FIXATION

$$N_2 \rightarrow NH_4^+$$

Nitrogen fixation is the process wherein N_2 is converted to ammonium, essential because it is the only way that organisms can attain nitrogen directly from the atmosphere. Certain bacteria, for example those among the genus *Rhizobium*, are the only organisms that fix nitrogen through metabolic processes.

Nitrogen fixing bacteria often form symbiotic relationships with host plants. Thissymbiosis is well-known to occur in the legume family of plants (e.g. beans, peas, and clover). In this relationship, nitrogen fixing bacteria inhabit legume root nodules and receive carbohydrates and a favorable environment from their host plant in exchange for some of the nitrogen they fix. There are also nitrogen fixing bacteria that exist without plant hosts, known as free-living nitrogen fixers. In aquatic environments, blue-green algae (really a bacteria called cyanobacteria) is an important free-living nitrogen fixer.

In addition to nitrogen fixing bacteria, high-energy natural events such as lightning, forest fires, and even hot lava flows can cause the fixation of smaller, but significant amounts of nitrogen. The high energy of these natural phenomena can break the triple bonds of N_2 molecules, thereby making individual N atoms available for chemical transformation.

Within the last century, humans have become as important a source of fixed nitrogen as all natural sources combined.

Burning fossil fuels, using synthetic nitrogen fertilizers, and cultivation of legumes all fix nitrogen. Through these activities, humans have more than doubled the amount of fixed nitrogen is pumped into the biosphere every year, the consequences of which are discussed below.

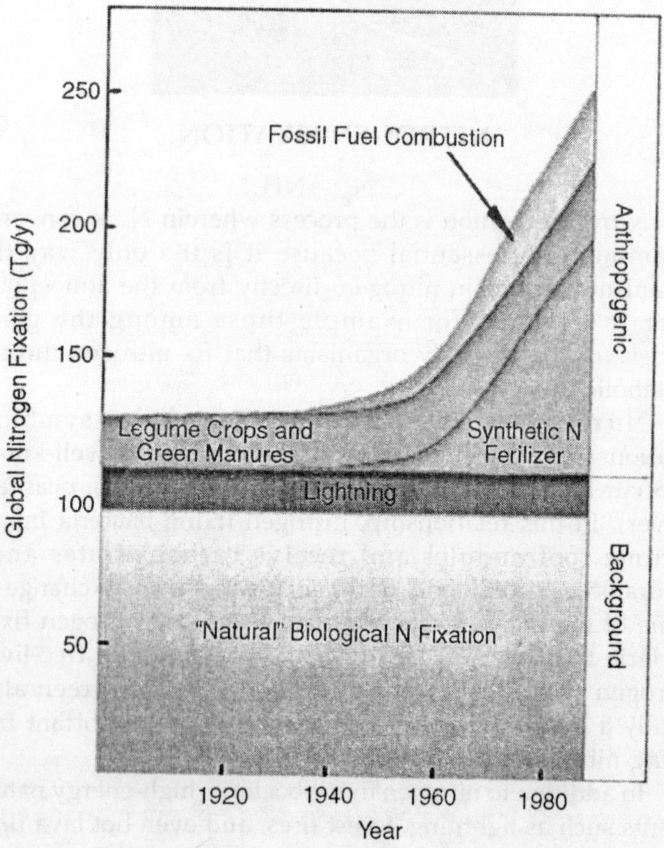

NITROGEN UPTAKE

$$NH_4^+ \rightarrow \text{Organic N}$$

The ammonia produced by nitrogen fixing bacteria is usually quickly incorporated into protein and other organic nitrogen compounds, either by a host plant, the bacteria itself, or another soil organism. When organisms nearer the top of the food chain (like us!) eat, we are using nitrogen that has been fixed initially by nitrogen fixing bacteria.

NITROGEN MINERALIZATION

$$\text{Organic N} \rightarrow NH_4^+$$

After nitrogen is incorporated into organic matter, it is often converted back into inorganic nitrogen by a process called nitrogen mineralization, otherwise known as decay. When organisms die, decomposers (such as bacteria and fungi) consume the organic matter and lead to the process of decomposition.

During this process, a significant amount of the nitrogen contained within the dead organism is converted to ammonium. Once in the form of ammonium, nitrogen is available for use by plants or for further transformation into nitrate (NO_3^-) through the process called nitrification.

NITRIFICATION

$$NH_4^+ \rightarrow NO_3^-$$

Some of the ammonium produced by decomposition is converted to nitrate via a process called nitrification. The bacteria that carry out this reaction gain energyfrom it. Nitrification requires the presence of oxygen, so nitrification can happen only in oxygen-rich environments like circulating or flowing waters and the very surface layers of soils and sediments.

The process of nitrification has some important consequences. Ammonium ions are positively charged and therefore stick (are sorbed) to negatively charged clay particles and soil organic matter. The positive charge prevents ammonium nitrogen from being washed out of the soil (or leached) by rainfall. In contrast, the negatively charged nitrate ion is not held by soil particles and so can be washed down the soil profile, leading to decreased soil fertility and nitrate enrichment of downstream surface and groundwater.

DENITRIFICATION

$$NO_3^- \rightarrow N_2 + N_2O$$

Through denitrification, oxidized forms of nitrogen such as nitrate and nitrite (NO_2^-) are converted to dinitrogen (N_2) and, to a lesser extent, nitrous oxide gas. Denitrification is an anaerobic process that is carried out by denitrifying bacteria, which convert nitrate to dinitrogen in the following sequence:

$$NO_3^- \rightarrow NO_2^- \rightarrow NO \rightarrow N_2O \rightarrow N_2.$$

Nitric oxide and nitrous oxide are both environmentally important gases. Nitric oxide (NO) contributes to smog, and nitrous oxide (N_2O) is an important greenhouse gas, thereby contributing to global climate change.

Once converted to dinitrogen, nitrogen is unlikely to be reconverted to a biologically available form because it is a gas and is rapidly lost to the atmosphere. Denitrification is the only nitrogen transformation that removes nitrogen from ecosystems (essentially irreversibly), and it roughly balances the amount of nitrogen fixed by the nitrogen fixers described above.

HUMAN ALTERATION OF N CYCLE

Early in the 20th century, a German scientist named Fritz Haber figured out how to short circuit the nitrogen cycle by fixing nitrogen chemically at high temperatures and pressures, creating fertilizers that could be added directly to soil. This technology has spread rapidly over the past century, and, along with the advent of new crop varieties, the use of synthetic nitrogen fertilizers has led to an enormous boom in agricultural productivity.

This agricultural productivity has helped us to feed a rapidly growing world population, but the increase in nitrogen fixation has had some negative consequences as well. While the consequences are perhaps not as obvious as an increase in global temperatures or a hole in the ozone layer, they are just as serious and potentially harmful for humans and other organisms. Not all of the nitrogen fertilizer applied to agricultural fields stays to nourish crops.

Some is washed off of agricultural fields by rain or irrigation water, where it leaches into surface or ground water and can accumulate.

In groundwater that is used as a drinking water source, excess nitrogen can lead to cancer in humans and respiratory distress in infants.

The U.S. Environmental Protection Agency has established a standard for nitrogen in drinking water of 10 mg per liter nitrate-N. Unfortunately, many systems (particularly in agricultural areas) already exceed this level.

By comparison, nitrate levels in waters that have not been altered by human activity are rarely greater than 1 mg/L. In surface waters, added nitrogen can lead to nutrient over-enrichment, particularly in coastal waters receiving the inflow from polluted rivers. This nutrient over-enrichment, also called eutrophication, has been blamed for increased frequencies of coastal fish-kill events, increased frequencies of harmful algal blooms, and species shifts within coastal ecosystems.

Reactive nitrogen (like NO_3^- and NH_4^+) present in surface waters and soils, can also enter the atmosphere as the smog-component nitric oxide (NO) and the greenhouse nitrous oxide (N_2O). Eventually, this atmospheric nitrogen can be blown into nitrogen-sensitive terrestrial environments, causing long-term changes. For example, nitrogen oxides comprise a significant portion of the acidity in acid rain which has been blamed for forest death and decline in parts of Europe and the Northeast United States.

Increases in atmospheric nitrogen deposition have also been blamed for more subtle shifts in dominant species and ecosystem function in some forest and grassland ecosystems. For example, on nitrogen-poor serpentine soils of northern Californian grasslands, plant assemblages have historically been limited to native species that can survive without a lot of nitrogen.

There is now some evidence that elevated levels of atmospheric N input from nearby industrial and agricultural development have paved the way for invasion by non-native plants. As noted earlier, NO is also a major factor in the formation of smog, which is known to cause respiratory illnesses like asthma in both children and adults.

Currently, much research is devoted to understanding the effects of nitrogen enrichment in the air, groundwater, and surface water. Scientists are also exploring alternative agricultural practices that will sustain high productivity while decreasing the negative impacts caused by fertilizer use.

These studies not only help us quantify how humans have altered the natural world, but increase our understanding of the processes involved in the nitrogen cycle as a whole.

FIXATION OF NITROGEN IN SOIL

NITROGEN IN BIOSPHERE

The growth of all organisms depends on the availability of mineral nutrients, and none is more important than nitrogen, which is required in large amounts as an essential component of proteins, nucleic acids and other cellular constituents. There is an abundant supply of nitrogen in the earth's atmosphere - nearly 79% in the form of N_2 gas.

However, N_2 is unavailable for use by most organisms because there is a triple bond between the two nitrogen atoms, making the molecule almost inert. In order for nitrogen to be used for growth it must be "fixed" (combined) in the form of ammonium (NH_4) or nitrate (NO_3) ions.

The weathering of rocks releases these ions so slowly that it has a neglible effect on the availability of fixed nitrogen. So, nitrogen is often the limiting factor for growth and biomass production in all environments where there is suitable climate and availability of water to support life.

Microorganisms have a central role in almost all aspects of nitrogen availability and thus for life support on earth:

- Some bacteria can convert N_2 into ammonia by the process termed nitrogen fixation; these bacteria are either free-living or form symbiotic associations with plants or other organisms (e.g. termites, protozoa)

- Other bacteria bring about transformations of ammonia to nitrate, and of nitrate to N_2 or other nitrogen gases
- Many bacteria and fungi degrade organic matter, releasing fixed nitrogen for reuse by other organisms.

All these processes contribute to the nitrogen cycle.

We shall deal first with the process of nitrogen fixation and the nitrogen-fixing organisms, then consider the microbial processes involved in the cycling of nitrogen in the biosphere.

EXPLANATION

A relatively small amount of ammonia is produced by lightning. Some ammonia also is produced industrially by the Haber-Bosch process, using an iron-based catalyst, very high pressures and fairly high temperature. But the major conversion of N_2 into ammonia, and thence into proteins, is achieved by microorganisms in the process called nitrogen fixation (or dinitrogen fixation).

The table below shows some estimates of the amount of nitrogen fixed on a global scale. The total biological nitrogen fixation is estimated to be twice as much as the total nitrogen fixation by non-biological processes.

Type of fixation 10^6 metric Non-biological	**N_2 fixed (10^{12} g per year, or tons per year)**
Industrial	about 50
Combustion	about 20
Lightning	about 10
Total	about 80
Biological	
Agricultural land	about 90
Forest and non-agricultural land	about 50
Sea	about 35
Total	about 175

To illustrate the importance of biological nitrogen fixation, the image below shows part of the Lower Sonoran desert in Arizona. Every plant that we see in this scene depends ultimately on biological nitrogen fixation.

Both free-living cyanobacteria and the cyanobacterial associates of lichens initially contributed nitrogen to the soil by forming a cryptobiotic crust.

Now numerous leguminous plants occur in this desert, with nitrogen-fixing *Rhizobium* in their root nodules. Examples are the green-stemmed brush-like trees at the right and left of the image (*Parkinsonia* species, common name "paloverde"), and several acacias and mesquites (not seen in this image).

MECHANISM

Biological nitrogen fixation can be represented by the following equation, in which two moles of ammonia are produced from one mole of nitrogen gas, at the expense of 16 moles of ATP and a supply of electrons and protons (hydrogen ions):

$$N_2 + 8H+ + 8e^- + 16\ ATP = 2NH_3 + H_2 + 16ADP + 16\ Pi$$

This reaction is performed exclusively by prokaryotes (the bacteria and related organisms), using an enzyme complex termednitrogenase. This enzyme consists of two proteins - an iron protein and a molybdenum-iron protein, as shown below.

The reactions occur while N_2 is bound to the nitrogenase enzyme complex. The Fe protein is first reduced by electrons donated by ferredoxin.

Then the reduced Fe protein binds ATP and reduces the molybdenum-iron protein, which donates electrons to N_2, producing HN=NH. In two further cycles of this process (each requiring electrons donated by ferredoxin) HN=NH is reduced to $H_2N\text{–}NH_2$, and this in turn is reduced to $2NH_3$.

Depending on the type of microorganism, the reduced ferredoxin which supplies electrons for this process is generated by photosynthesis, respiration or fermentation.

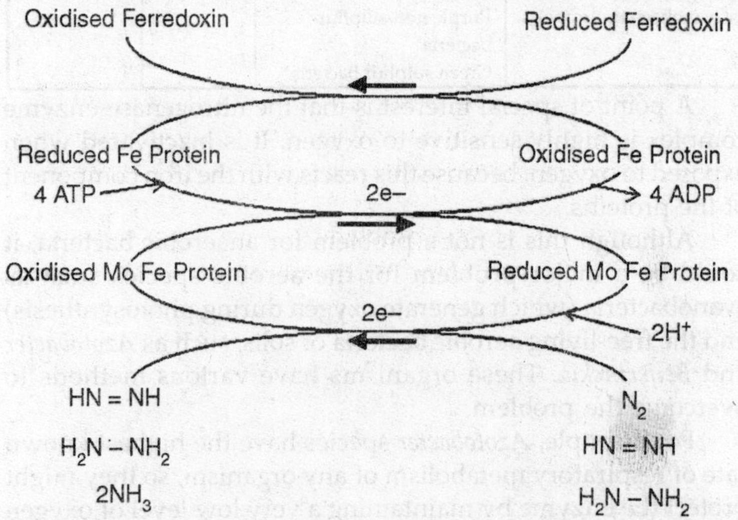

There is a remarkable degree of functional conservation between the nitrogenase proteins of all nitrogen-fixing bacteria.

The Fe protein and the Mo-Fe protein have been isolated from many of these bacteria, and nitrogen fixation can be shown to occur in cell-free systems in a laboratory when the Fe protein of one species is mixed with the Mo-Fe protein of another bacterium, even if the species are very distantly related.

FIXING ORGANISMS

All the nitrogen-fixing organisms are prokaryotes (bacteria). Some of them live independently of other organisms - the so-called free-living nitrogen-fixing bacteria. Others live in intimate symbiotic associations with plants or with other organisms (e.g. protozoa). Examples are shown in the table below.

Table. Examples of Nitrogen-fixing Bacteria
(* Denotes a Photosynthetic Bacterium)

Free living		Symbiotic with plants	
Aerobic	Anaerobic Winogradsky column for details)	Legumes	Other plants
Azotobacter *Beijerinckia* *Klebsiella* (some) Cyanobacteria (some)*	*Clostridium* (some) *Desulfovibrio* Purple sulphur bacteria* Purple non-sulphur bacteria* Green sulphur bacteria*	*Rhizobium*	*Frankia* *Azospirillum*

A point of special interest is that the nitrogenase enzyme complex is highly sensitive to oxygen. It is inactivated when exposed to oxygen, because this reacts with the iron component of the proteins.

Although this is not a problem for anaerobic bacteria, it could be a major problem for the aerobic species such as cyanobacteria (which generate oxygen during photosynthesis) and the free-living aerobic bacteria of soils, such as *Azotobacter* and *Beijerinckia*. These organisms have various methods to overcome the problem.

For example, *Azotobacter* species have the highest known rate of respiratory metabolism of any organism, so they might protect the enzyme by maintaining a very low level of oxygen in their cells. *Azotobacter* species also produce copious amounts of extracellular polysaccharide.

By maintaining water within the polysaccharide slime layer, these bacteria can limit the diffusion rate of oxygen to the cells. In the symbiotic nitrogen-fixing organisms such as *Rhizobium*, the root nodules can contain oxygen-scavenging molecules such as leghaemoglobin, which shows as a pink colour when the active nitrogen-fixing nodules of legume roots are cut open.

Leghaemoglobin may regulate the supply of oxygen to the nodule tissues in the same way as haemoglobin regulates the supply of oxygen to mammalian tissues. Some of the cyanobacteria have yet another mechanism for protecting nitrogenase: nitrogen fixation occurs in special cells (heterocysts) which possess only photosystem (used to generate ATP by light-mediated reactions) whereas the other cells have both photosystem (which generates oxygen when light energy is used to split water to supply H_2 for synthesis of organic compounds).

SYMBIOTIC NITROGEN FIXATION

Legume Symbioses

The most familiar examples of nitrogen-fixing symbioses are the root nodules of legumes (peas, beans, clover, etc.).

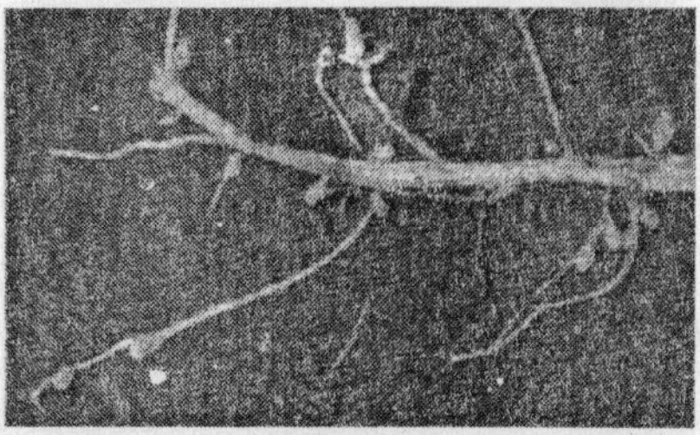

Part of a clover root system bearing naturally occurring nodules of *Rhizobium*. Each nodule is about 2-3 mm long.

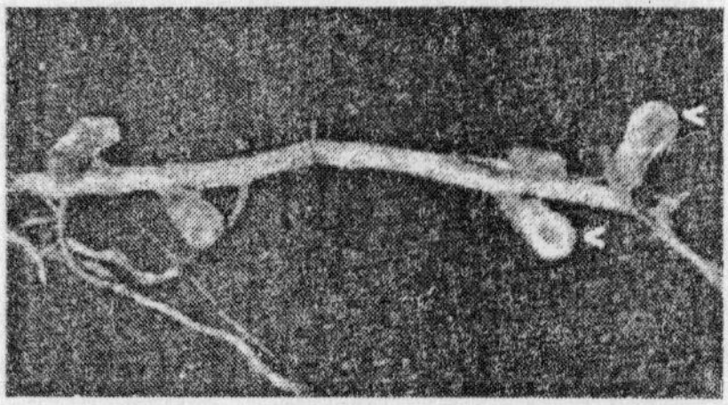

Clover root nodules at higher magnification, showing two partly crushed nodules (arrowheads) with pink-coloured contents.

This colour is caused by the presence of the pigment leghaemoglobin - a unique metabolite of this type of symbiosis. Leghaemoglobin is found only in the nodules and is not produced by either the bacterium or the plant when grown alone.

In these leguminous associations the bacteria usually are *Rhizobium* species, but the root nodules of soybeans, chickpea and some other legumes are formed by small-celled rhizobia termed *Bradyrhizobium*. Nodules on some tropical leguminous plants are formed by yet other genera.

In all cases the bacteria "invade" the plant and cause the formation of a nodule by inducing localised proliferation of the plant host cells.

Yet the bacteria always remain separated from the host cytoplasm by being enclosed in a membrane - a necessary feature in symbioses.

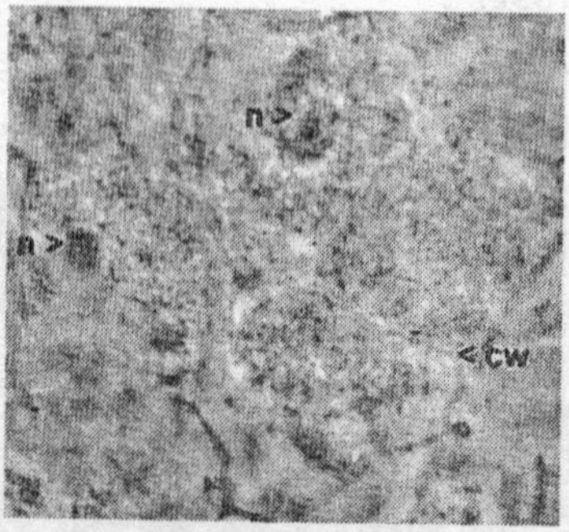

Part of a crushed root nodule of a pea plant, showing four root cells containing colonies of *Rhizobium*. The nuclei (n) of two root cells are shown; cw indicates the cell wall that separates two plant cells. Although it cannot be seen clearly in this image, the bacteria occu r in clusters which are enclosed in membranes, separating them from the cytoplasm of the plant cells.

In nodules where nitrogen-fixation is occurring, the plant tissues contain the oxygen-scavenging molecule, leghaemoglobin(serving the same function as the oxygen-carrying haemoglobin in blood). The function of this molecule in nodules is to reduce the amount of free oxygen, and thereby to protect the nitrogen-fixing enzyme nitrogenase, which is irreversibly inactivated by oxygen.

Associations with *Frankia*

Frankia is a genus of the bacterial group termed actinomycetes - filamentous bacteria that are noted for their production of air-borne spores. Included in this group are the common soil-dwelling *Streptomyces* species which produce many of the antibiotics used in medicine. *Frankia* species are slow-growing in culture, and require specialised media, suggesting that they are specialised symbionts.

They form nitrogen-fixing root nodules (sometimes called actinorhizae) with several woody plants of different families, such as alder (*Alnus* species), sea buckthorn (*Hippophae rhamnoides*, which is common in sand-dune environments) and *Casuarina* (a Mediterranean tree genus). Figure shows a young alder tree (*Alnus glutinosa*) growing in a plant pot, and Figure shows part of the root system of this tree, bearing the orange-yellow coloured nodules (arrowheads) containing *Frankia*.

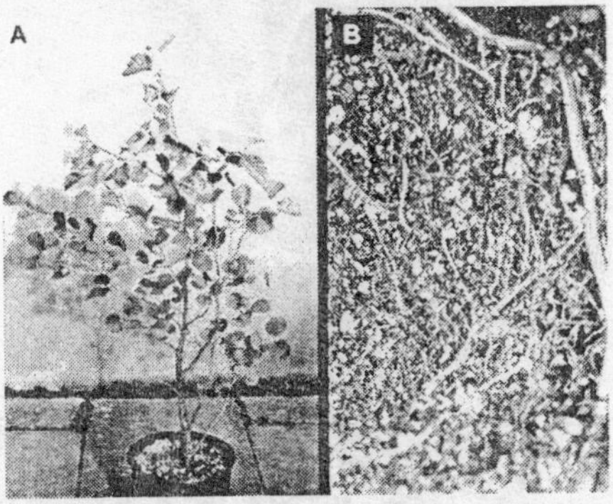

Alder and the other woody hosts of *Frankia* are typical pioneer species that invade nutrient-poor soils. These plants probably benefit from the nitrogen-fixing association, while supplying the bacterial symbiont with photosynthetic products.

Cyanobacterial Associations

The photosynthetic cyanobacteria often live as free-living organisms in pioneer habitats such as desert soils (cyanobacteria) or as symbionts with lichens in other pioneer habitats.

They also form symbiotic associations with other organisms such as the water fern *Azolla*, and cycads.The association with *Azolla*, where cyanobacteria (*Anabaena azollae*) are harboured in the leaves, has sometimes been shown to be important for nitrogen inputs in rice paddies, especially if the fern is allowed to grow and then ploughed into the soil to release nitrogen before the rice crop is sown.

A symbiotic association of cyanobacteria with cycads is shown below. The first image shows a pot-grown plant. The second image shows a close-up of the soil surface in this pot.

Short, club-shaped, branching roots have grown into the aerial environment. These aerial roots contain a nitrogen-fixing cyanobacterial symbiont.

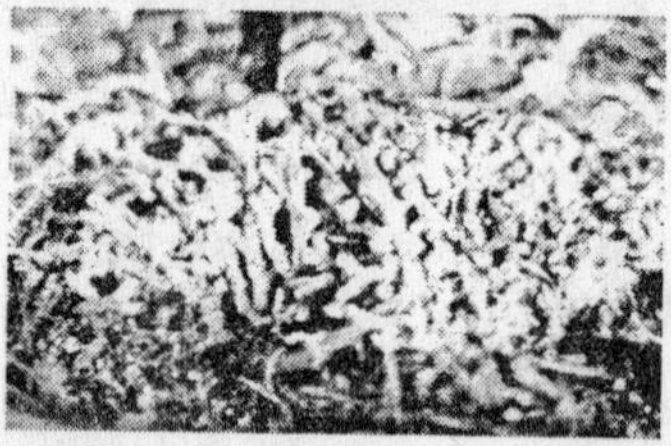

In addition to these intimate and specialised symbiotic associations, there are several free-living nitrogen-fixing bacteria that grow in close association with plants. For example, *Azospirillum* species have been shown to fix nitrogen when growing in the root zone (rhizosphere) or tropical grasses, and even of maize plants in field conditions.

Similarly, *Azotobacter* species can fix nitrogen in the rhizosphere of several plants. In both cases the bacteria grow at the expense of sugars and other nutrients that leak from the roots. However, these bacteria can make only a small contribution to the nitrogen nutrition of the plant, because nitrogen-fixation is an energy-expensive process, and large amounts of organic nutrients are not continuously available to microbes in the rhizosphere.

This limitation may not apply to the bacteria that live in root nodules or other intimate symbiotic associations with plants. It has been estimated that nitrogen fixation in the nodules of clover roots or other leguminous plants may consume as much as 20% of the total photosynthate.

He diagram below shows an overview of the nitrogen cycle in soil or aquatic environments. At any one time a large proportion of the total fixed nitrogen will be locked up in the biomass or in the dead remains of organisms (shown collectively as "organic matter"). So, the only nitrogen available to support new growth will be that which is supplied by nitrogen fixation from the atmosphere or by the release of ammonium or simple organic nitrogen compounds through the decomposition of organic matter. Some of other stages in this cycle are mediated by specialised groups of microorganisms and are explained below.

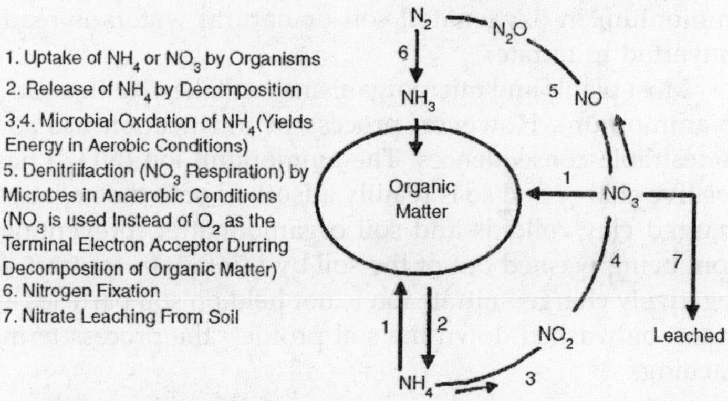

NITRIFICATION

The term nitrification refers to the conversion of ammonium to nitrate (pathway 3-4). This is brought about by the nitrifying bacteria, which are specialised to gain their energy by oxidising ammonium, while using CO_2 as their source of carbon to synthesise organic compounds. Organisms of this sort are termed chemoautotrophs - they gain their energy by chemical oxidations (chemo-) and they are autotrophs (self-feeders) because they do not depend on pre-formed organic matter.

In principle the oxidation of ammonium by these bacteria is no different from the way in which humans gain energy by oxidising sugars. Their use of CO_2 to produce organic matter is no different in principle from the behaviour of plants.

The nitrifying bacteria are found in most soils and waters of moderate pH, but are not active in highly acidic soils. They almost always are found as mixed-species communities (termed consortia) because some of them - e.g. *Nitrosomonas* species - are specialised to convert ammonium to nitrite (NO_2^-) while others - e.g. *Nitrobacter* species - convert nitrite to nitrate (NO_3^-).

In fact, the accumulation of nitrite inhibits *Nitrosomonas,* so it depends on *Nitrobacter* to convert this to nitrate, whereas *Nitrobacter* depends on *Nitrosomonas* to generate nitrite. The nitrifying bacteria have some important environmental consequences, because they are so common that most of the ammonium in oxygenated soil or natural waters is readily converted to nitrate.

Most plants and microorganisms can take up either nitrate or ammonium. However, process of nitrification has some undesirable consequences. The ammonium ion (NH_4^+) has a positive charge and so is readily adsorbed onto the negatively charged clay colloids and soil organic matter, preventing it from being washed out of the soil by rainfall. In contrast, the negatively charged nitrate ion is not held on soil particles and so can be washed down the soil profile - the process termed leaching.

In this way, valuable nitrogen can be lost from the soil, reducing the soil fertility. The nitrates can then accumulate in groundwater, and ultimately in drinking water. There are strict regulations governing the amount of nitrate that can be present in drinking water, because nitrates can be reduced to highly reactive nitrites by microorganisms in the anaerobic conditions of the gut. Nitrites are absorbed from the gut and bind to haemoglobin, reducing its oxygen-carrying capacity. In young babies this can lead to respiratory distress - the condition known as "blue baby syndrome". Nitrite in the gut also can react with amino compounds, forming highly carcinogenic nitrosamines.

Denitrification

Denitrification refers to the process in which nitrate is converted to gaseous compounds (nitric oxide, nitrous oxide and N_2) by microorganisms. The sequence usually involves the production of nitrite (NO_2^-) as an intermediate step. Several types of bacteria perform this conversion when growing on organic matter in anaerobic conditions. Because of the lack of oxygen for normal aerobic respiration, they use nitrate in place of oxygen as the terminal electron acceptor.

This is termed anaerobic respiration and can be illustrated as follows: In aerobic respiration (as in humans), organic molecules are oxidised to obtain energy, while oxygen is reduced to water:

$$C_6H_{12}O_6 + 6\ O_2 = 6\ CO_2 + 6\ H_2O + \text{energy}$$

In the absence of oxygen, any reducible substance such as nitrate (NO_3^-) could serve the same role and be reduced to nitrite, nitric oxide, nitrous oxide or N_2.

Thus, the conditions in which we find denitrifying organisms are characterised by a supply of oxidisable organic matter, and absence of oxygen but availability of reducible nitrogen sources. A mixture of gaseous nitrogen products is often produced because of the stepwise use of nitrate, nitrite, nitric oxide and nitrous oxide as electron acceptors in anaerobic respiration. The common denitrifying bacteria include several species of *Pseudomonas, Alkaligenes* and *Bacillus*. Their activities result in substantial losses of nitrogen into the atmosphere, roughly balancing the amount of nitrogen fixation that occurs each year.

IMPORTANCE

Molecular nitrogen (N_2) is the major component (approximately 80%) of the earth's atmosphere. The element nitrogen is an essential part of many of the chemical compounds, such as proteins and nucleic acids, which are the basis of all life forms. However, N_2 cannot be used directly by biological systems to build the chemicals required for growth and reproduction. Before its incorporation into a living system, N_2 must first be combined with the element hydrogen. This process of reduction of N_2, commonly referred to as "nitrogen fixation" (N-fixation) may be accomplished chemically or biologically.

Nitrogen (N) is the nutrient element most frequently found limiting to the growth of green plants. This results from the continual loss of nitrogen from the reserve of combined or fixed nitrogen, which is present in soil and available for use by plants. It is continually depleted by such processes as microbial denitrification, soil erosion, leaching, chemical volatilization, and perhaps most important, removal of nitrogen-containing crop residues from the land.

The nitrogen reserve of agricultural soils must therefore be replenished periodically in order to maintain an adequate (non-growth limiting) level for crop production. This replacement of soil nitrogen is generally accomplished by the addition of chemically fixed nitrogen in the form of commercial inorganic fertilizers or by the activity of biological nitrogen fixation (BNF) systems. The significance of BNF as the major mechanism of recycling of nitrogen from the unavailable atmospheric form to available forms in the biosphere cannot be overemphasized.

MECHANISM

Atmospheric nitrogen (N) is a molecule composed of two atoms of nitrogen linked by a very strong triple bond. Large amounts of energy are required to break this bond and the molecule is therefore quite chemically unreactive. The general chemical reaction for the fixation of nitrogen:

$$(N + 3H_2 + \text{Energy} \rightarrow 2NH_3)$$

is identical for both the chemical and the biological processes. The triple bond of N must be broken and three atoms of hydrogen must be added to each of the nitrogen atoms. Living organisms use energy derived from the oxidation ("burning") of carbohydrates to reduce molecular nitrogen (N_2) to ammonia (NH_3). The chemical process of nitrogen fixation involves "burning" of fossil fuels to obtain the electrons, hydrogen atoms and energy needed to reduce molecular nitrogen.

ENERGY COST FOR NITROGEN FIXATION

The reduction of nitrogen, whether accomplished chemically or biologically, requires a large amount of energy. The chemical process used to produce fertilizers utilizes vast amounts of fossil fuels as an energy source.

These materials are nonreplaceable and, ultimately, exhaustible. BNF obtains the required energy from the oxidation of carbohydrates which have been formed by the photosynthetic activity of green plants. The energy for photosynthesis comes from sunlight. The energy for BNF is therefore indirectly derived from a universally available and inexhaustible source. The direct source of energy (carbohydrate) for BNF is therefore potentially available wherever conditions permit the growth of photosynthetic organisms.

The ability of a biological system to fix nitrogen is dependent on the presence of a particular enzyme system known as nitrogenase, which catalyzes the conversion of N_2 into a reduced form (ammonia combined with certain organic compounds) which can then be used for growth by microorganisms and higher life forms.

The nitrogenase system consists of two different protein molecules (enzymes) which must function together in the nitrogen fixing process. One of the enzymes (azoferredoxin) is an iron-containing protein.

The second enzyme (molybdoferredoxin) contains both iron and molybdenum. The two components combine and function together as a single system. Components of the nitrogenase system from different organisms have been combined to form an "active hybrid" nitrogenase system, indicating an apparent general similarity in BNF systems derived from different organisms. Nitrogenase has so far been detected only in certain microorganisms.

The ammonia resulting from fixation is rapidly incorporated into certain amino acids, such as glutamine or alanine. The nitrogen may then be transferred to other amino acids and nitrogen-containing compounds by a variety of commonly occurring amino transfer reactions.

MAGNITUDE OF BNF

It is estimated that BNF on a global scale may reach a value of 175 million metric tons of nitrogen fixed per year. The amount of nitrogen fixed in any given situation would depend upon the environmental conditions and the nature of biological system(s) present which are capable of nitrogen fixation. For particular situations, nitrogen fixation rates may vary from barely detectable to several hundred kilograms per hectare per year. The significance of the contribution of any BNF system to the nitrogen economy in any situation is a function of the supply and demand of the biological community for nitrogen.

DIVERSITY OF BNF SYSTEMS

BNF is known to occur to a varying degree in many different environments, including soils; fresh and salt waters and sediments; on or within the roots, stems, and leaves of certain higher plants; and within the digestive tracts of some animals. The potential for nitrogen fixation exists for any environment capable of supporting growth of microorganisms. Biological systems which are capable of fixing nitrogen are historically classified as no symbiotic or symbiotic, depending on the required involvement of one or more than one organism, respectively, in the process.

SYMBIOTIC NITROGEN FIXATION

The most important contribution to BNF comes from the symbiotic association of certain micro-organisms with the roots of higher plants. A classic example is that of the bacteria (Rhizobium) which characteristically infect the roots of leguminous plants (e.g., bean, soybean, clover, peanut) with a high degree of host specificity. Small nodules are formed on the roots and these become filled with an altered form of the bacteria (bacteroides) which fix appreciable amounts of nitrogen. This symbiosis alone accounts for 20% of global biological nitrogen fixed annually. The legumes represent a major direct source of food for man and forage for livestock and therefore represent a critical contribution to world food production.

It is significant that agricultural scientists have learned to manipulate this symbiotic relationship in agronomic practice, employing selected combinations of bacteria and legumes in specific situations to obtain maximum crop production on land which is of low fertility and frequently unsuitable for growth of non-legume crops.

Nitrogen fixation rates from 75 to 300 kilograms of N per hectare per year (70 to 275 pounds of N per acre per year) are common in various combinations. In Florida, the growth of white clover in combination with a tropical forage grass may result in the fixation of 180 kg N/ha/yr (160 lb N/a/yr) under favorable conditions.

Numerous genera of non-leguminous angiosperms, such as Alnus, Casuarina, Coriaria, Myrica, etc., form root nodules in response to infection by the actinomycete Frankia. These associations may achieve fixation rates as high as 100 kg N/ha/yr and may occur as climax vegetation or as pioneer species in adverse soil environments. Some gymnosperms, such as Cycas, Macrozamia, and Podocarpus, are capable of forming similar nitrogen fixing root nodule associations.

A variety of additional plant-symbiotic nitrogen fixing associations have been reported. Examples include the bacterium Kiebsiella in leaf nodules of Psychotria, and associations of cyanobacteria with fungi (lichens), liverworts (Blasia), angiosperms (Gunnera) and the water fern Azolla.

Increasing knowledge concerning the genetic information in various microorganisms which confers nitrogen-fixing ability now make the "creation" of new and perhaps more efficient nitrogen fixing organisms and symbiotic associations a possibility.

NON-SYMBIOTIC NITROGEN FIXATION

There is great diversity in the metabolic types of free-living microorganisms which are capable of BNF. This includes about 20 genera of non-photosynthetic aerobic (Azotobacter, Beijerinckia) and anaerobic (Clostridium) bacteria and about 15 genera of photosynthetic cyanobacteria (blue-green algae) such as Anabaena and Nostoc.

Free-living, non-photosynthetic bacteria depend on soil organic matter as a food source whereas the photosynthetic microorganisms may derive their food from the products of photosynthesis.

The nitrogen fixing activity of free-living, non-photosynthetic, aerobic bacteria is strongly dependent on favorable moisture conditions, oxygen, and an organic food source. Anaerobic representatives (Clostridium) predominate in grassland and waterlogged soils and soil aggregates where moisture conditions and organic substrates are available but oxygen supply to the micro-environment of the bacteria is severely restricted.

The amounts of nitrogen fixed by free-living non-photosynthetic bacteria in the soil may achieve an approximate maximum of 15 kilograms per hectare per year. This relatively low estimated contribution is the result of limited availability of suitable organic substrates (energy sources) and low bacterial populations in the soil environment. Nitrogen fixation is characteristically higher in environments such as tropical soils, where such factors as substrate availability, temperature and moisture are more favorable to the maintenance and activity of a high bacterial population. Amendment of soil with a readily used organic substrate generally results in some increase in nitrogen fixation. Attempts to increase fixation in unamended soil by addition of high populations of bacteria (soil inoculation) are generally unsuccessful. The increased population of N-fixing bacteria resulting from inoculation is temporary and will rapidly die back to the original number found in an unamended soil, where no provision has been made to create environmental changes which will favour a higher microbial population.

Very little information is available concerning the possible contribution of the free-living, photosynthetic cyanobacteria to the nitrogen economy of soils but maximum gains of fifty kilograms per hectare per year have been reported. Nitrogen-fixing activity of these organisms is, of course strongly dependent on adequate sunlight in addition to favorable moisture conditions.

Living plant roots release a wide variety of simple organic compounds which may be used as food by free-living soil bacteria. This continuous supply of food supports a higher microbial population in the soil immediately surrounding the plant root (rhizosphere). Evidence indicates that native nitrogen-fixing bacteria are common in the rhizosphere of certain plants and that they may fix significant amounts of nitrogen in some cases.

This effect may be related to the closeness of the root-microorganism association. Food material released from the roots would be available in greater concentration to those microorganisms more closely associated with the root surface (Azospirillum). A striking example is seen in certain combinations of bacteria with some tropical grasses which have a high photosynthetic efficiency and grow under environmental conditions favoring high photosynthetic activity.

The roots of such plants may supply the nitrogen-fixing microorganisms with a relatively high and sustained supply of food (photosynthate) which is available in limited supply in the rhizosphere of most plants.

Under appropriate environmental conditions, aerobic cyanobacteria may contribute significantly to nitrogen gains in fresh water environments and anaerobic Clostridium may play a similar important role in fresh water sediments. Most of the nitrogen fixation in marine environments (about 20% of the total amount of annual biological fixation) is attributed to the cyanobacteria but many other kinds of nitrogen-fixing microorganisms have also been found in such environments.

The significant contribution of photosynthetic (cyanobacteria) and non-photosynthetic (Clostridium) microorganisms to nitrogen fixation in the rhizosphere of rice is well recognized. In addition, nitrogenase activity has been detected in the soil surrounding the mycorrhizal roots of several conifers and in the rhizospheres of marine (Thalassia) and freshwater (Glyceria) angiosperms.

The leaf surface (phyllosphere) of certain plants in warm, humid tropical regions may provide an additional favorable environment for the growth and nitrogen fixing activity of such free-living bacteria as Azotobacter, Beijerinckia and Kiebsiella.

Free-living nitrogen-fixing anaerobic bacteria are present and fix nitrogen in the intestinal contents of a variety of animals (herbivores) and also man. Nitrogenase activity is generally quite low and its significance in terms of satisfying nutritional requirements of the host appears doubtful.

FUNCTION OF SUPER PHOSPHATE IN PLANTS

ESSENTIAL ELEMENTS FOR PLANT GROWTH

The biological functions of P in living organisms is most notable in the ubiquitious ATP/ADP energy transport and storage compounds. Additionally, sugar phosphates form the "rails" of the nucleic acids DNA and RNA (which N-containing bases forming the "rungs"). Phospholipids are an important constituent of membrane chemistry and phosphoproteins are essential for life functions.

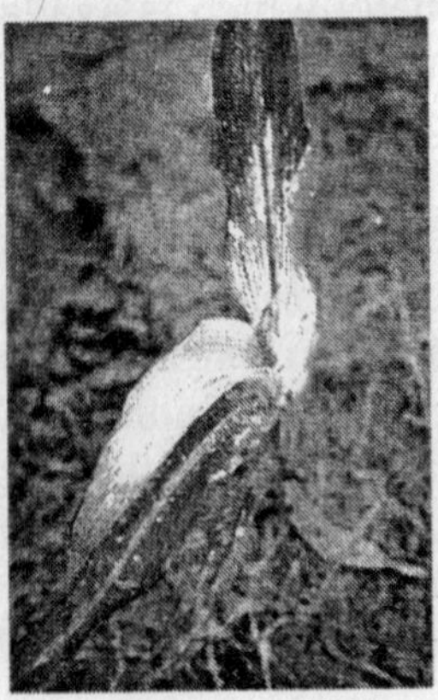

Phosphorus is phloem-mobile and the physiological results of P deficiency are spread more or less evenly around the plant, usually with no glaring visual deficiency symptom except for stunted growth and late maturity. Grassy species, including corn, will show reddening of leaves if P is severely deficient.

FORMS

The forms of P in soil are: soluble, mineral, and organic.

Soluble P

In soils in often in the micromolar concentration range and is an almost negligible per cent of total soil P except for the fact that it is the only P that plants can use! The forms of soluble P are $H_2PO_4^-$ and HPO_4^{2-}, which the proportion between them dependent upon pH. At pH 7.2, they are present in equal concentrations. At pH 6.2, the ratio $H_2PO_4^-:HPO_4^{2-}$ is 10:1, and at pH 8.2, 1:10.

Mineral P

Constitutes 20 to 85% of total soil P, depending upon the soil itself, particularly the soil organic matter content. Generally, formation of Al and Fe phosphates, either by precipitation or sorption, limits P solubility in acid soils and Ca phosphates limit P solubility in alkaline soils. This picture creates a "sweet spot" for maximum P solubility in equilibrium with P minerals at pH 6.5. By the same token, the minimum amount of P retention, or fixation, (within the range of common soil pH values) is at pH ~6.5.

Organic P

Constitute 15 to 80% of total P (and exactly complements the % mineral P since soluble P is numerically negligible.) Soil organic matter typically contains C/N/P/S in a ratio of 140:10:1.3:1.3, so that knowledge of the amount of organic matter will provide an estimate of the amount of organic P.

Organo-P is cleaved by enzymes called phosphatases, which are present in many soil organisms and produced by some plants as well. In some soils, P is the most limiting plant nutrient, not N. In such soils, usually in the tropics, the pH is in the acid range, soils have a great P-fixing capacity, soil organic matter levels are very low, and no organic residues are returned to soils.

Another way of looking at P, particularly from a management point of view, is: soluble-P ⇔ labile-P ⇔ nonlabile-P

Soluble P is fairly well defined, of course, but the chemistry of the labile P is more complicated, being a combination of mineral and organic P, and both absorbed and precipitated.

In addition, the mineralization of organic-P is a particularly valuable source of bio available P, particularly in strongly-P-fixing soils. The soluble P may be thought of as an intensity and labile P as a quantity, so that the P buffer capacity is quantity/intensity.

P Inputs to soil

- Plant and animal residues
- Fertilizers
- (Aeolian dust)

P Outputs from soil

- Crop removal
- Runoff/erosion (mainly loss of top soil, the particulates of which contain organic-P)
- (Leaching to subsoil or groundwater is of little or no consequence)

P TESTING

The most common methods of testing for P availability are the Bray-1, Mehlich III, and Olsen methods. Bray 1 uses a mixture of HCl and NH_4F to dissolve adsorbed and poorly crystallized phosphate minerals into an acid extractant that increases the solubility of phosphate for purposes of extractant; fluoride is a good complexing agent for Al^{3+} to improve phosphate solubility. The Mehlich method is similar and uses a mixture of NH_4F, acetic acid, and NH_4Cl/HCl. Both are applicable for measuring labile phosphate in acid and neutral soils. For calcareous soils (containing free $CaCO_3$, with pH>7.5) such acid extractants would simply dissolve the carbonates and consume the acid in the extractant.

Instead, the Olsen method uses 0.5 M $NaHCO_3$ at pH 8.5, which suppresses Ca^{2+} by both the high HCO_3^- concentration and high pH, allowing phosphates to dissolve out of calcium phosphate minerals (by the common ion principle). (All of these extractants differ from direct measure of the soluble phosphate in soil solution, which is present in analytically-difficult, but not impossibly low, concentrations and is only a small portion of the total plant-available phosphate in soil.)

Yet another P-test that is becoming increasingly popular is iron-oxide impregnated filter paper, which uses to advantage the strong sorption of phosphate on iron oxides. A solution of ferric chloride is neutralized to form iron oxides on filter paper.

The iron-oxide impregnated filter paper can either be shaken with soil and water or buried in soil for a period of time to absorb soil phosphate and then recovered. The phosphate is then desorbed from the filter paper and analysed. Results generally show a strong correlation between labile P by this method and by the more conventional methods.

Results of these analyses are used with tables of low, medium, high and very high values, based on probability and degree of plant response to added P. Such tables are collected soil-by-soil and crop-by-crop using field trials and may be distributed also as computer-computed recommendations.

PHOSPHORUS FERTILIZER ANALYSIS

Fertilizer analyses of phosphorus in the U.S. is in the form of P_2O_5.

Remember: To convert %P to %P_2O_5, multiply by,

$$(2 \times 31 + 5 \times 16)/2 \times 31\ [= 2.29].$$

Analyses are divided into water-soluble P_2O_5 and citrate-soluble P_2O_5. Monocalcium phosphate, the major portion of superphosphate, is water-soluble. Dicalcium phosphate is largely citrate-soluble. Rock phosphate has low citrate-soluble P and very low water-soluble P.

P Fertilizers

Rock Phosphate

Source of all mineral P for fertilizers. Ore grade contains 6 to 15% P. Mined from deposits in Florida, Tennessee, North Carolina, Morocco, and elsewhere. Very insoluble. Effectively used as fertilizer material only on very acid soils or if acidulated or partially acidulated.

Single Superphosphate

Aka, normal, simple, or ordinary superphosphate (although how can anything "super" be both normal or ordinary?) Basically the product of the reaction of rock phosphate, usually a fluorapatite, with sulfuric acid according to the reaction:

$$2\ Ca_5(PO_4)_3F + 7\ H_2SO_4 \rightarrow 3\ Ca(H_2PO_4)_2 + 7\ CaSO_4 + 2\ HF$$

Producing a mixture of monocalcium phosphate, a water-soluble phosphate, and gypsum, along with the volatile, and very nasty, hydrofluoric acid gas as a by-product. (Superphosphate plants are regulated with regard to HF and SiF_4 emissions.) The commercial product contains between 7 to 9.5% P, or 16 to 22% P_2O_5.

Phosphoric Acid, H_3PO_4

Produced by the same process as single superphosphate, but using more sulfuric acid.

$$2\ Ca_5(PO_4)_3F + 10\ H_2SO_4 \rightarrow 6\ H_3PO_4 + 10\ CaSO_4 + 2\ HF$$

Phosphoric acid is then separated mechanically from the gypsum for use. Phosphoric acid is rarely used directly as a fertilizer except for purposes of fertigation and commercial hydroponics, particularly when the source water contains (calcium) bicarbonate alkalinity (=high pH) that would otherwise cause phosphate precipitation.

This material is, however, the acid used to manufacture the ammonium phosphates and the next fertilizer: Pure phosphoric acid is a solid that melts at 42°C. Phosphoric acid of commerce is typically ~23%P, or 53-54% 16 to 22% P_2O_5.

Triple Superphosphate

The active ingredient is calcium monophosphate, as in the single superphosphate, but the product carries less gypsum content because it was made with phosphoric acid instead of sulfuric acid:

$$2\ Ca_5(PO_4)_3F + 12\ H_3PO_4 + 9\ H_2O \rightarrow 9\ Ca(H_2PO_4)_2 + 2\ HF$$

This material contains 17 to 23% P, 44 to 52% P_2O_5. It was the most popular P fertilizer until surpassed in usage by ammonium phosphates. Much of the gypsum generated as a byproduct is diverted into construction.

Ammonium Phosphates

MAP, monoammonium phosphate, $NH_4H_2PO_4$, typically 12-52-0, and DAP, diammonium phosphate, $(NH_4)_2HPO_4$, typically 20-50-0.

Formulated by reacting various quantities of ammonia with phosphoric acid. More often regarded as P fertilizer than N fertilizer since fertilizer application of DAP to achieve desired N rates exceeds usual P rates, in most cases. Completely water-soluble P.

Superphosphoric Scid

A phosphoric acid mixture containing a substantial proportion of one or more polyphosphoric acids (pyrophosphoric acid, $H_4P_2O_7$; tripyrophosphoric acid, $H_5P_3O_{10}$; tetrapyrophosphoric acid, $H_6P_4O_{13}$. Manufactured by "roasting" phosphoric acid to remove some water, thereby polymerizing two or more phosphates. Typically 30 to 36% P, or 68 to 83% P_2O_5

Ammonium Polyphosphates, APP

Product of reacting polyphosphoric acid, with ammonia

to form the final product, with a typical fertilizer composition of 10-34-0, 10% N and 15% P. The product is very water-soluble, and has the capability to chelate some micronutrients, such as Fe and Zn.

Organic Sources of P

The very same sources of organic N mentioned previously, for example, manure, whey, sludge, etc., also contain organic P that can be serve as a P source as the organic matter mineralizes. Typical values are 4 lbs P/1000 gal in (solid) manure and whey and 6.3 lbs P/1000 gal in sewage sludge. The use of organic sources as both N and P sources is sometimes problematic because the amount required to meet N needs may also add P far in excess of crop needs; the extra P in the top soil may increase the hazard of P runoff to surface waters.

Other so-called "organic" P sources are bonemeal, in which the active P ingredient is closely related to a carbonatic rock phosphate, and rock phosphate (which is not really organic at all). Both are generally acceptable to organic farmers and gardeners. The availability of P in both is increased by acidulation, either partial or complete, to convert the apatite minerals to more soluble phosphates, although this process is frowned upon by organic farmers because of its use of harsh chemicals in the process.

Guano, bird and bat dung accumulations, are also popular as organic fertilizer but wreak a heavy cost to the environments from which they are extracted. This outcome should be no surprise considering the history of the guano trade, as recounted in the Trade Environment Database.

PLANT NUTRIENTS

Sixteen chemical elements are known to be important to a plant's growth and survival. The sixteen chemical elements are divided into two main groups: non-mineral and mineral

Non-Mineral Nutrients

The Non-Mineral Nutrients are hydrogen (H), oxygen (O),

& carbon (C). These nutrients are found in the air and water.

In a process called photosynthesis, plants use energy from the sun to change carbon dioxide (CO_2 - carbon and oxygen) and water (H_2O- hydrogen and oxygen) into starches and sugars. These starches and sugars are the plant's food.

Photosynthesis means "making things with light". Since plants get carbon, hydrogen, and oxygen from the air and water, there is little farmers and gardeners can do to control how much of these nutrients a plant can use.

The 13 mineral nutrients, which come from the soil, are dissolved in water and absorbed through a plant's roots. There are not always enough of these nutrients in the soil for a plant to grow healthy. This is why many farmers and gardeners use fertilizers to add the nutrients to the soil.

The mineral nutrients are divided into two groups: macronutrients and micronutrients.

Macronutrients

Macronutrients can be broken into two more groups: primary and secondary nutrients. The primary nutrients are nitrogen (N), phosphorus (P), and potassium (K). These major nutrients usually are lacking from the soil first because plants use large amounts for their growth and survival.

The secondary nutrients are calcium (Ca), magnesium (Mg), and sulfur (S). There are usually enough of these nutrients in the soil so fertilization is not always needed. Also, large amounts of Calcium and Magnesium are added when lime is applied to acidic soils. Sulfur is usually found in sufficient amounts from the slow decomposition of soil organic matter, an important reason for not throwing out grass clippings and leaves

Micronutrients are those elements essential for plant growth which are needed in only very small (micro) quantities. These elements are sometimes called minor elements or trace elements, but use of the term micronutrient is encouraged by the American Society of Agronomy and the Soil Science Society of America. The micronutrients are boron (B), copper (Cu), iron(Fe), chloride (Cl), manganese (Mn), molybdenum (Mo) and zinc(Zn). Recycling organic matter such as grass clippings and tree leaves is an excellent way of providing micronutrients (as well as macronutrients) to growing plants

SOIL

In general, most plants grow by absorbing nutrients from the soil. Their ability to do this depends on the nature of the soil. Depending on its location, a soil contains some combination of sand, silt, clay, and organic matter.

The makeup of a soil (soil texture) and its acidity (pH) determine the extent to which nutrients are available to plants Soil Texture (the amount of sand, silt, clay, and organic matter in the soil) Soil texture affects how well nutrients and water are retained in the soil.

Clays and organic soils hold nutrients and water much better than sandy soils. As water drains from sandy soils, it often carries nutrients along with it. This condition is called leaching. When nutrients leach into the soil, they are not available for plants to use.

An ideal soil contains equivalent portions of sand, silt, clay, and organic matter. Soils across North Carolina vary in their texture and nutrient content, which makes some soils more productive than others. Sometimes, the nutrients that plants need occur naturally in the soil. Othertimes, they must be added to the soil as lime or fertilizer.

Chapter 4

Importance of Elements and Compounds

In a university course of inorganic chemistry a balance has to be struck between a uniform treatment of all of the elements, and a greater concentration on the socially more important ones. The question of which are the socially more important elements is to some extent a matter of opinion.

A chemist working in the plastics industry would doubtless come up with a different list from one working in the metals industry, and someone working with agricultural chemicals would doubtless have different ideas from someone working on ceramics.

ABUNDANCE OF DIFFERENT ELEMENTS

The percentages are percentages by mass - the percentage numbers of atoms would of course be different. All abundances over 0.1% have been included. The figures reveal that some 99% of the matter in the world around us is made up of as few as 14 elements. Note, however, that many of the remaining elements in the biosphere, despite their low concentration, are essential to its healthy continuance.

Atmosphere		Hydrosphere	
Nitrogen	75.5%	Oxygen	85.8%
Oxygen	23.0%	Hydrogen	10.7%
Argon	1.4%	Chlorine	2.1%
		Sodium	1.1%
Total	**99.9%**	**Total**	**99.7%**

Lithosphere		Biosphere	
Oxygen	46.6%	Oxygen	52.4%
Silicon	27.7%	Carbon	39.4%
Aluminium	8.1%	Hydrogen	6.6%
Iron	5.0%	Nitrogen	0.5%
Calcium	3.6%	Calcium	0.4%
Sodium	2.8%	Potassium	0.2%
Potassium	2.6%		
Magnesium	2.1%		
Titanium	0.4%		
Total	**98.9%**	**Total**	**99.5%**

CONSUMPTION OF DIFFERENT ELEMENTS

Siliceous materials and limestone used for building purposes have been excluded, and a lower limit has been set at 0.1 million tons.

Millions of tons	Per year
Carbon	1000-10000
Iron, sodium	100-1000
Calcium, nitrogen, oxygen, potassium, sulfur	10-100
Aluminium, barium, chlorine, chromium, copper, fluorine, hydrogen, lead, magnesium, manganese, phosphorus, zinc	1-10
Boron, nickel, tin, titanium, zirconium	0.1-1

Of the 25 elements in the table, 10 are very plentiful, Of the remainder, the majority occur in relatively concentrated deposits, from which they or their compounds can be extracted fairly easily (e.g. hydrocarbons from petroleum).

For copper, lead, zinc, nickel, and tin, demand is such that less concentrated deposits have to be worked, down to 0.5% in the case of copper and 0.2% in the case of tin.

ELEMENTS FROM VERY LOW-GRADE ORES

The following table gives the minimum metal content that

an ore must have before it can be mined and smelted economically with present-day technology. Values are given up to a limit of 0.5%, and may be compared with the value for iron of 25%.

Platinum	0.0003%
Gold	0.001%
Silver	0.01%
Uranium	0.2%
Mercury	0.2%
Tin	0.2%
Molybdenum	0.3%
Copper	0.5%

CLASSIFICATION OF ELEMENTS

TYPES OF ELEMENT

The classification of elements into metals and nonmetals, with the intermediate category of semimetal, is a fundamental one in inorganic chemistry. *Metallic elements* are characterized by the following properties under ordinary conditions:

- High lustre, high opacity over the visible spectrum;
- High electrical and thermal conductivity.

Nonmetals are characterized by:

- No lustre, high transparency over the visible spectrum;
- Low electrical and thermal conductivity.

Semimetals are characterized by:

- High opacity over the visible spectrum, little or no lustre;
- Intermediate electrical conductivity.

"High opacity over the visible spectrum" is a way of saying that the substance absorbs light strongly over the whole range of the visible spectrum, or at least over the whole range beyond the red. If it is not lustrous, therefore, the substance appears black to the eye, or very dark brown.

"High transparency over the visible spectrum" is a way of saying that the substance absorbs light weakly over the greater part of the visible spectrum, if not over all of it. If no light is absorbed, the substance appears colourless or white. (Colourless substances look white when finely divided because light is scattered from the many crystal faces. Similarly, coloured substances become paler on grinding.) If some light is absorbed, the substance appears coloured, the colour being the complementary one to the colour of the light that is absorbed.

ELECTRICAL CONDUCTIVITY

$$\kappa = 1/\rho$$

The resistivity is defined by the equation:

$$R = \rho(l/A)$$

where R is the resistance of a uniform conductor, l is its length, and A is its cross sectional area. The SI unit of ρ is ohm meter (Ω m) and its value is equal to the resistance that a meter cube of the substance offers to the passage of electricity from one parallel face to another.

The SI unit of k is thus W^{-1} m^{-1}.On the basis of their ability to transmit an electric current, materials can be divided into four main classes: *Ordinary conductors*, characterized by having conductivities greater than about 10^5 $\Omega^{-1}m^{-1}$ at room temperature, which fall with increase of temperature.

Semiconductors, characterized by having conductivities between about 10^{-7} and 10^5 $\Omega^{-1}m^{-1}$ at room temperature, which rise with increase of temperature, and are highly sensitive to traces of impurity.

Electrolytic conductors, characterized by the chemical decomposition that takes place at the points at which a direct current enters and leaves the material.

A further characteristic of metals that is usually given is that, in the solid state, they are malleable (easily hammered into sheets) and ductile (easily drawn into wires). By contrast, nonmetals in the solid state are brittle and easily powdered. In practice, however, while the majority of elements that are metallic according to the above criteria are indeed malleable and ductile, there are a few that do not have this property, even when ultra-pure. (Slight traces of impurity can produce brittleness in metals. Thus, ordinary commercial tungsten is so brittle that it can only be worked with difficulty, whereas the ultra-pure metal can be cut with a hacksaw, turned, drawn, or extruded.) Since the exceptional elements include some whose metallicity has never bee questioned - e.g. manganese, cobalt, and zinc - it seems better *not* to make malleability and ductility a characteristic of metals, but rather a property possessed by most of them.

Zinc is a particularly interesting case, because it is malleable at some temperatures and brittle at others. Thus, it is brittle at room temperature, but softens above 100 ºC. At 205 ºC, however, it becomes brittle again, and can then be powdered in a mortar.

ASSIGNMENT OF ELEMENTS TO CLASSES

The clear-cut nonmetals are: hydrogen, nitrogen, oxygen, fluorine, sulfur, chlorine, bromine, and the inert or noble gases (helium, neon, argon, krypton, xenon). The clear-cut semimetals are: boron, silicon, germanium, tellurium. The clear-cut metals constitute all of the remaining elements, with the exception of the problem cases discussed below.

CARBON

Under ordinary conditions, carbon exists in two common forms, graphite and diamond, of which graphite is the more stable. Diamond is definitely nonmetallic, but graphite has properties on the borderline between a semimetal and a metal. Thus it is a shiny, black solid, readily separated into flakes, with a metallic conductivity in the plane of the flakes, and a semimetallic conductivity perpendicular to the plane. Other forms include black, petrol-soluble "fullerenes". When pure, these are insulators.

PHOSPHORUS

This exists in a number of different forms: one white,

several red, and several black. The white form can be made by condensing the vapour. It melts at 44 °C and boils at 280 °C. A red form is obtained when the white form is heated to just below its boiling point. A black form can be obtained by heating the white form under a very high pressure, or in the presence of a mercury catalyst. The yellow and red forms are insulators; the black forms are semiconductors. The most stable form under ordinary conditions is black; all forms, however, melt to a colourless liquid.

ARSENIC

This can be obtained in several forms, ranging from a yellow one, which is similar to white phosphorus, to a grey one, which is a metal. The metallic form is the most stable under ordinary conditions; the yellow one is very unstable.

ANTIMONY

This can also be obtained in several forms, ranging from yellow to metallic. The metallic form is again the most stable, with the yellow very unstable.

SELENIUM

This exists in three red forms, one black form, and one grey. The red and black forms are insulators; the grey is a semiconductor. The latter is the most stable.

IODINE

Iodine crystallizes in blackish-grey opaque scales, with an almost metallic lustre. In the plane of the scales it has the conductivity of a semiconductor; perpendicular to the plane it is an insulator.

VOLATILITIES OF METALS AND NONMETALS

Metals are generally involatile, i.e. they have boiling points in excess of 500 ºC. Only mercury has a boiling point below 500 ºC, this being at 357 ºC. Most metals have their boiling point somewhere between 1000 ºC and 4000 ºC. The highest value is for tungsten, at 5660 ºC.

Nonmetals vary from being very volatile to very involatile. All the definite nonmetals listed above have boiling points below 500 ºC, and most are gases at room temperature. The least volatile is sulfur, boiling at 445 ºC. If, however, we include carbon among the nonmetals, as surely we should, this lies at the opposite extreme of volatility, boiling at 4830 ºC.

CHEMICAL PROPERTIES

There are many chemical differences between metals and nonmetals.

Some elements have very similar properties and are grouped into families.

- *These include*:
- *Alkali metals*: Li, Na, K, Rb, Cs
- *Alkaline earth metals*: (Mg,) Ca, Sr, Ba
- *Halogens*: F, Cl, Br, I
- *Inert or noble gases*: He, Ne, Ar, Kr, Xe

"Alkali" is an Arabic word meaning "the ash" (ash from plants contains sodium and potassium carbonates).

"Halogen" means "producer-of (-gen) salt (Greek *hals*)".

Some chemists class magnesium as an alkaline earth metal as it is quite similar to calcium, often occurring with it in nature.

The gases He, Ne, Ar, Kr, and Xe were called "inert" when they were thought to be completely inactive. Their name was changed to "noble" after the xenon fluorides were discovered. This was because metals having a low activity (but nevertheless some activity) like gold and platinum are sometimes called "noble" as they are used for noble purposes. The change was inept. "Inert" need not imply complete inactivity, and "noble" is hardly apt (commoners breathe in these gases no less than nobles!).

CLASSIFICATION OF COMPOUNDS

Compounds can be classified according to the number of different elements that they contain: binary (two), ternary (three), etc.

TYPES OF BINARY COMPOUND

Corresponding to the two limiting types of element, metal and nonmetal, there are three limiting types of binary compound:

- *Metallic*: These are formed principally by the combination of a metal with a metal, and have the characteristics of a metal.
- *Nonmetallic*: These are formed principally by the combination of a nonmetal with a nonmetal, and have the characteristics of a nonmetal.
- *Salt-like*: These are formed principally by the combination of a metal with a nonmetal, and constitute a new class of material.

Metallic Compounds

An example is the compound Mg_2Sn. This is made by fusing together a mixture of magnesium and tin of the appropriate composition, and cooling the resulting melt. The product is homogeneous, as can be seen by cutting and polishing it, and examining the polished surface under a microscope.

If more magnesium or tin is used than required by the formula Mg_2Sn, the product ceases to be homogeneous - crystals of Mg_2Sn can be seen embedded in a matrix of smaller crystals containing the excess of magnesium or tin. The compound Mg_2Sn is a bluish-white lustrous solid, having all the properties of a metal.

Metal-metal systems vary in character between the following extremes:

- Those in which the two metals are able to form solid solutions with each other over the whole composition range from 100% of one to 100% of the other. An example of such a system is that between silver and gold.
- Those in which the two metals are completely immiscible with each other, even in the liquid state. An example of such a pair of metals is iron and lead.

- Those in which a number of compounds are formed.

Metallic compounds fall into one of two categories of compound:

- *Stoicheiometric compounds*: These are compounds that obey the law of constant composition to very high accuracy, e.g. CH_4, NaCl, Mg_2Sn. They are also called "definite" compounds or "Daltonides". [The word "stoicheiometric" comes from "stoicheiometry", the name given to the measurement of the proportions of elements in compounds]
- *Non-stoicheiometric compounds*: These have a variable composition within certain limits. They are also called "indefinite" compounds or "Berthollides".

  Examples of non-stoicheiometric compounds are provided by the copper-tin system.

The three compounds referred to above have the following compositions at 300 ºC $CuZn_{0.85-0.95}$, $CuZn_{1.4-2.0}$, and $CuZn_{2.5-6.2}$.

The same compositions are obtained at room temperature if the solids are cooled fairly quickly. The first compound, however, is metastable at room temperature, and when cooled very slowly, decomposes at about 250 ºC into a mixture of the second compound and $CuZn_{0-0.5}$, a solid solution of zinc in copper (brass).

The formulae of compounds of this type are normally written in terms of some kind of idealized formula, using a *circa* ("round about") sign ~. The choice of idealized formula is based upon structural or theoretical considerations, and is not without arbitrariness. In the case of copper-zinc compounds, the idealized formulae are CuZn, Cu_5Zn_8, and $CuZn_3$, and the compounds are written ~CuZn, ~Cu_5Zn_8, and ~$CuZn_3$.

NONMETALLIC COMPOUNDS

Pure silica also forms colourless crystals, melting to a colourless liquid at about 1600 ºC and boiling at about 2400 ºC. Both compounds are insulators. Conductivity measurements have also been made on liquid silica, in which state it remains a poor conductor.Nonmetallic compounds are generally formed by the combination of a nonmetal with a nonmetal, but combinations of a metal and a nonmetal can also be nonmetallic if the influence of the nonmetal predominates.

This occurs:

- When the ratio of metal atoms to nonmetal atoms is relatively low,
- When the metallic character of the metal is relatively weak, or
- When both circumstances pertain.

Good examples of nonmetallic compounds of metals are provided by the highest oxides of chromium and manganese. These have the formulae CrO_3 and Mn_2O_7, and are also similar to the highest oxides of sulfur and chlorine, SO_3 and Cl_2O_7. Thus CrO_3 is a red solid melting at 196 ºC, and Mn_2O_7 is a dark red oil. Further, electrical measurements on CrO_3 show this to be an insulator in both the solid and liquid state. Both compounds decompose below their boiling point, Mn_2O_7 explosively.

Salt-Like Compounds

These represent a new class, characterized by the following properties:

- No lustre, high transparency over the visible spectrum,
- Either
  - Low electrical conductivity in the solid state, but high conductivity in the molten state and in aqueous solution, conduction being of the electrolytic type, or
  - High electrolytic conductivity in both the solid and the liquid state.

PSEUDO-BINARY COMPOUNDS

A great many compounds of ternary or higher order behave as if they were compounds of binary order, with a group of atoms acting as if it was a single atom. Such a group is called a "radical". A classical example is provided by the cyanides, some of which are described below:

Hydrogen cyanide, HCN. This is manufactured by passing a mixture of methane and ammonia over a platinum catalyst at 1200 °C:

$$CH_4 + NH_3 \rightarrow HCN + 3H_2$$

It is a colourless liquid, boiling at 26 °C to a colourless gas. It is a poor conductor of electricity. Like most cyanides, it is extremely poisonous.

Sodium cyanide, NaCN: This is manufactured by absorbing gaseous hydrogen cyanide in aqueous sodium hydroxide or sodium carbonate. The compound is a colourless solid, melting at 564 ºC to a colourless liquid. The liquid conducts electricity with the formation of sodium at the cathode and a colourless gas at the anode. This gas is cyanogen, C_2N_2.

Potassium cyanide, KCN: This is similar to sodium cyanide.

Cyanogen, C_2N_2. This is a colourless gas, condensing at – 21 ºC to a colourless liquid.

Cyanogen chloride, ClCN: This is made by passing chlorine into aqueous hydrogen cyanide:

$$HCN + Cl_2 \rightarrow HCl + ClCN$$

It is a colourless gas, condensing at 13 ºC to a colourless, mobile liquid. From the above descriptions, it is clear that there is a close resemblance between cyanides and halides, both in composition and in character:

Composition		**Character**
Cyanide	*Chloride*	
HCN	HCl	Nonmetallic
NaCN	NaCl	Salt-like
KCN	KCl	Salt-like
ClCN	ClCN	Nonmetallic

This resemblance extends to the parent substances:

$(CN)_2$ Cl_2 Nonmetallic

So close is this resemblance that cyanogen is called a pseudo-halide. A second example is provided by ammonium compounds, e.g. ammonium chloride, NH_4Cl. These are very similar to alkali metal compounds.

Indeed, ammonium compounds and potassium compounds often form mixed crystals with each other.

Furthermore, electrolysis of ammonium chloride solution with a mercury cathode at 0 ºC gives a grey mass that is similar in character to sodium or potassium amalgam. Thus it reduces copper dichloride solution to copper, zinc chloride to zinc, etc.

On warming it decomposes into ammonia and hydrogen in the proportions required by the equation,

$$2NH_4(\text{amalgam}) \rightarrow 2NH_3 + H_2$$

the NH_4 radical is thus very much like an alkali metal atom in its behaviour.

The nature of pseudo-binary compounds: The nature of pseudo-binary compounds follows from our discussion of binary compounds. The bonding between radicals and radicals or radicals and atoms is the same as the bonding between atoms and atoms.

For example:

- The molecules HCN, ClCN, and $(CN)_2$ are held together with essentially covalent bonds:

  H–CN, Cl–CN, NC–CN

- The compounds NaCN and KCN are ionic:

  $Na^+[CN]$, $K^+[CN]^-$

- Ammonium chloride is ionic: $[NH_4]^+Cl^-$d.
- Ammonium amalgam presumably contains:

  $[NH_4]^+e^-$

Bonding *within* radicals is of a covalent character. Thus the full bond formula of the HCN molecule is H–CºN while that of the $[CN]^-$ ion is $^-$CºN Ions like $[CN]^-$ and $[NH_4]^+$ are called "molecular" ions. The square brackets indicate that the charge is somewhere on the radical as a whole, but they are often omitted:

$$CN^- \text{ and } NH_4^+.$$

This should not be misinterpreted as implying, for example, that the charge on the cyanide ion is located on the nitrogen atom.

RADICALS

Formula of radical	Name as electropositive constituent of compounds	Name as electronegative constituent of compounds

Non-metal-like

Table.Forming one Covalent Bond or a Singly Charged Anion

OH		Hydroxide
NO_2	Nitryl	Nitrite
NO_3		Nitrate
CN	cyanogen	Cyanide
SCN	thiocyanogen	Thiocyanate
CH_3CO_2		Acetate
CH_3 (Me)	methyl	Methyl
C_6H_5 (Ph)	phenyl	Phenyl

Table. Forming two Covalent Bonds or a Doubly Charged Anion

CO_3		Carbonate
SO_3		Sulfite
SO_4		Sulfate

Table.Forming three Covalent Bonds or a Triply Charged Anion

PO_4		Phosphate

Metal-like

Table. Forming a singly charged cation

NH_4	Ammonium
NMe_4	Tetramethylammonium

- All the nonmetal-like radicals form covalent bonds with sufficiently electronegative atoms or radicals. This includes nitrate and sulfate, e.g. in $MeNO_3$ and Me_2SO_4.

- All the nonmetal-like radicals form anions with sufficiently electropositive elements. This includes methyl and phenyl, e.g. in Na^+Me^- and Na^+Ph^-.
- The NMe_4 radical forms a similar amalgam to NH_4.

NOMENCLATURE OF RADICALS

The names given above are based on those recommended by IUPAC – the International Union of Pure and Applied Chemistry Though not very systematic, they reflect long established usage.

The use of the endings –ite and –ate to indicate a lower and higher oxo-anion (e.g. nitrite, nitrate) parallels the use of –ous and –ic to indicate a lower and higher cation. British schools teach different names as advised by the Association for Science Education (ASE):

IUPAC	ASE
Nitrite	Nitrate(III)
Nitrate	nitrate(V)
Sulfite	Sulfate(IV)
Sulfate	Sulfate(VI)
Acetate	Ethanoate

The ASE names are derived from IUPAC rules for naming less common ions. These names, however, are only used in British schools. Everywhere else in the world (in the chemical literature, chemical industry, other sciences, and other countries) IUPAC names are used.

NOMENCLATURE OF PSEUDO-BINARY COMPOUNDS

Pseudo-binary compounds are named in a similar way to binary compounds. In general, the name of the electropositive constituent is cited first, and that of the electronegative constituent second.

Strictly speaking, this requires the incorporation of radicals in the electrochemical series, but in practice this is only done very roughly. Thus NO_3 and SO_4 are always regarded as being the electronegative constituent of a compound, even though in a few cases this is disputable, e.g. in FNO_3.

Examples of nomenclature:

CH_3Cl	Methyl chloride
$NaCH_3$	Sodium methyl
$(CH_3)_2SO_4$	Dimethyl sulfate
FNO_3	Fluorine nitrate

Involatile Nonmetallic Compounds

Some nonmetallic compounds are involatile because their molecules are very large (e.g. higher paraffins). Although the van der Waals interactions between atoms in different molecules are small (they are the same as the interactions between atoms of small molecules), there are so many atoms that the total interaction between molecules is large.

Compounds of this type nevertheless dissolve in volatile solvents (e.g. higher paraffins in lower ones), because each molecule can interact with as many solvent molecules as it can get round it.

These interactions add up to overcome the large interaction between molecules of the compound. Very large molecules are called "macromolecules".

Other involatile nonmetallic compounds are different in being insoluble in volatile solvents (e.g. diamond, silicon carbide, and silica).

The nature of these compounds can be inferred as follows:

- From the above discussion of volatile compounds, the molecules of these compounds in the vapour phase must interact very strongly in the liquid and solid phases.
- In many volatile compounds, carbon atoms are known to bond to four other atoms in directions defined by a tetrahedron. Now if carbon atoms were to bond to each other in this kind of way, instead of a small molecule being produced, a continuous three-dimensional network would be\ formed, each atom being held in the network by very strong bonds -

the same kind of bonds that hold the atoms of each molecule together in volatile compounds.

In two dimensions this may be drawn:

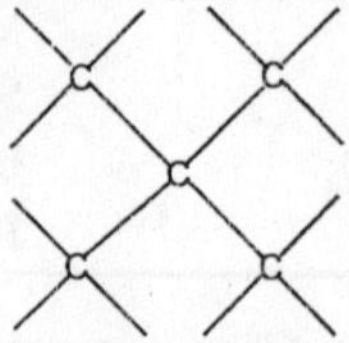

Such a structure would satisfy:

- And suggests itself for diamond.
- A determination of the structure of diamond by X-ray crystallography shows that it indeed has this structure, with carbon-carbon bond lengths of 1.54 Å, as compared with 1.53 Å in C_2H_6.
- In a similar way we can "guess" the structures of silicon carbide and silica. Silicon also forms volatile compounds in which the silicon atom is surrounded by a tetrahedron of other atoms (e.g. simple molecules like SiH_4 and $SiCl_4$, and resolvable ones of the type SiWXYZ). Silicon carbide could thus have a similar structure to that of diamond:

Further, since oxygen forms compounds in which its atom has two neighbours, silica could have a structure of the diamond type, only with an oxygen atom between every pair of silicon atoms.

That silicon carbide and silica indeed have structures of this kind has been shown by X-ray crystallography. (Both compounds are polymorphic, but in each case the different forms have the same arrangement of nearest neighbours, and differ only in the relative orientation of more distant neighbours.)

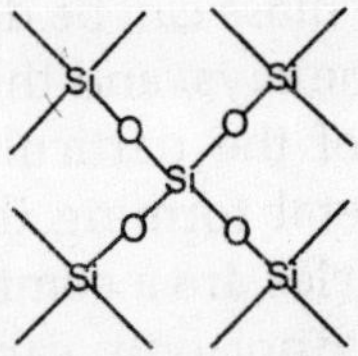

Compounds of this type thus consist of giant assemblies of atoms tightly held together. Such assemblies of atoms are called "frameworks" and such compounds "nonmolecular". Liquefaction of compounds of this type involves the breaking of some of the bonds, for only in this way can atoms move positions.

Clearly, however, the system will resist whole-scale breaking of bonds, and fluidity will rely on a process whereby the breaking of one bond is compensated for by the remaking of another bond somewhere else. Consequently the fluidity of such compounds in the liquid state is low: that is, their viscosity is high.

METALLIC COMPOUNDS

The nature of these is more difficult to determine. However, we can get so far along the following line:

- Metals conduct electricity without electrolysis taking place. The carriers of electricity cannot therefore be ions. They must either be some kind of charged entity that can travel through a conductor without affecting it or else a component of the atoms making up the conductor, which, by a process of successive displacement from one atom to another, can effectively be transferred from one end of the conductor to the other.
- When an electric current is passed through a tube of gas at low pressure, rays can be seen to emerge from the negative electrode (the cathode). These rays are drawn towards a plate carrying a positive charge, and must therefore be carrying negative charge. They are likewise deflected by a magnet. From measurement of these deflections, the ratio of the

charge to the mass can be determined for the particles making up the rays, and this is found to be the same, irrespective of the particular gas in the tube or the particular metal forming the cathode. This suggests that the particles are a common constituent of matter, and that they not only carry electricity through the gas, but also through the metal that supplies the electricity to the gas, the particles streaming without change from one into the other.

- Measurement of the charge on these particles gives a value identical in magnitude with the charge on the sodium ion worked out above, viz. F/L. This gives us our terminology for the particle ("electron") and mode of conduction ("electronic"): it also gives us something else. It suggests that metal atoms may be thought of as being composed of metal ions and electrons, sodium atoms being Na^+ with one electron, magnesium atoms Mg^{2+} with two, and so on.
- A simple model of a metal, therefore, is that of a collection of metal ions, held together by the appropriate number of electrons situated between them, the electrons being readily made to move by supplying further electrons at one point of the metal and taking electrons away at another. In the solid, the electrons hold the ions together in a rigid assembly: in the liquid the ions are free to move.
- The constitutional formulae of metals can thus be written in the same way as the constitutional formulae of salts. The latter are properly written as exemplified by $(Na^+Cl^-)_n$, with n a very large number, but this is usually abbreviated to Na^+Cl^-, that is, to one unit of the complete assembly. In an analogous way, metallic sodium can be written Na^+e^-, metallic magnesium $Mg^{2+}(e^-)_2$, etc. A metallic compound like Cs_7O may similarly be written $(Cs^+)_7(e^-)_5O^{2-}$.
- Support for these formulations is provided by the latent heats of vaporization of metals:

	ΔH
Sodium, Na^+e^-	90 kJ mol^{-1}
Magnesium, $Mg^{2+}(e^-)_2$	130 kJ mol^{-1}
Aluminium, $Al^{3+}(e^-)_3$	290 kJ mol^{-1}

These values are for the vaporization of the liquid at the boiling point into gaseous atoms. It can be seen that the strength of binding increases as the number of opposite charges increases.

The vapours of metals are, as we have seen, quite different from the solids and liquids. In general, metals vaporize into a mixture of atoms and diatomic molecules, the proportion of the latter diminishing with increase of temperature.

Thus sodium metal just above the boiling point contains about 99% of sodium atoms and 1% of Na_2 molecules. On vaporization of a metal, therefore, the individual ions or pairs of ions become separated from each other, carrying their electrons with them. The mechanism for the conduction of electricity accordingly breaks down, and metallic properties disappear.

METALLICITY

The degree of metallic character of a bond in a binary compound is best measured by the electrical conductivity of the compound. Strictly speaking, the property is not of an isolated bond (as in Na_2) but of bonds in concert.

TYPES OF FORMULA

Empirical formulae: The empirical formula of a compound is the simplest possible formula indicating the relative numbers of different atoms in it. The formula is determined from the composition of the compound.

Molecular Formulae

The molecular formula of a compound indicates the total numbers of different atoms in a molecule of it. For molecular compounds, the molecular formula is determined from the empirical formula and the molecular mass of the compound.

For nonmolecular compounds, it is set at the empirical formula multiplied by infinity.

In practice, a mixture of empirical and molecular formulae are used – NaCl alongside C_2H_6. Even hybrid formulae are used, e.g.

Na_2O_2 (as this compound contains O_2^{2-} ions). Ideally, the different formulae should be distinguished, e.g. with brackets:

(NaCl) $[C_2H_6]$ $(Na_2[O_2])$

Structural formulae: The structural formula of a compound indicates the way in which the atoms are united in the molecule.

There are two main types:

1. Constitutional formulae,
2. Descriptive formulae.

1. Constitutional formulae:

   a. *Covalent compounds*: *bond formulae*: These are determined from the valencies of the atoms and such structural information as is required.
   Example:

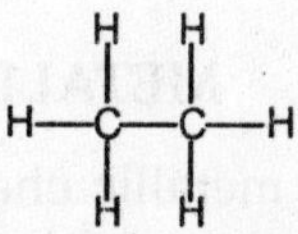

   b. *Ionic compounds*: *ion formulae* These are determined from the charges on the ions.
   Example: Na^+Cl^-

2. *Descriptive formulae*: These do not attempt to give any indication of the bonding.

   a. *Single stroke formulae*: These are formulae in which bonded atoms are connected up with a single stroke, with no indication of bond number or polarity. They are usually used for compounds in which the bonds have an intermediate bond number or polarity.
   An example is the following representation of one layer of graphite:
   Example:

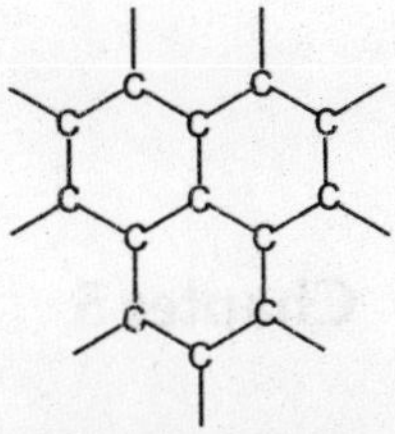

This kind of representation is also used in referring to bond distances in a compound, as in the statement "the C–C bond length in graphite is 1.42 Å". Care needs to be taken not to confuse formulae of this kind with bond formulae.

b. *Descriptive formulae of crystals*: If the crystal structure of a nonmolecular compound is known, the arrangement of atoms can be represented in the form $[AB_{x/y}]_{\infty}$, or more generally $[AB_{x/y}B'_{x'/y'}\ldots]_{\infty}$, where x is the number of nearest neighbours B to A, and y is the number of nearest neighbours A to B – $1/y$ being the share atom A has of each neighbouring atom B.
Example: $[NaCl_{6/6}]$

Chapter 5

Introduction to Periodic Table

ALKALI METALS

The alkali metals are the first group of the periodic table of the elements. These metals are lithium (Li), sodium (Na), potassium (K), rubidium (Rb), caesium (Cs) and francium (Fr). They have got one outer electron and the connection to the atomic nucleus is very weak.

They are the least noble substances. In the air they react very fast with oxygen and so they are covered in an oxide layer. Alkali metals are kept in liquid hydrocarbons like petroleum. The reactivity grows from lithium to caesium.

They react suddenly with water, then they burn or can even explode. Because of the big radius of caesium his outer electron is very free, so caesium reacts so strong with oxygen that it begins to burn in the air.

When these metals react with water the product is a caustic solution, like caustic soda solution (NaOH). They are very soft, too.

ALKALINE EARTH METALS

The alkaline earth metals are the second group of the periodic table of the elements. These metals are beryllium (Be), magnesium (Mg), calcium (Ca), strontium (Sr), barium (Ba) and radium (Ra). These elements have got two outer electrons and so they did not react as strong as the alkali metals, because also can make caustic solutions. These elements are much harder than the alkali metals. Magnesium is an element which do not only burns in oxygen, but in all substances which contain oxygen. So magnesium burns in water, too. Divers use magnesium flares to see something under water.

BORON GROUP

The boron group is the third group of the periodic table of the elements. These elements are boron (B), aluminium (Al), gallium (Ga), Indium (In) and thallium (Tl). Boron is a metalloid and there other four elements are metals. The most used element of this group is aluminium. They elements are reactively. When aluminium is used it is covered in an oxide layer.

CARBON SILICON GROUP

The carbon silicon group is the fourth group of the periodic table of the elements. These elements are carbon (C), silicon (Si), germanium (Ge), tin (Sn) and lead (Pb). Carbon is so noble that it is a nonmetal. Silicon and germanium are metalloids and tin and lead are metals. Lead is a very soft metal. Silicon is very often found in the compound silicon dioxide (quartz). Silicon and germanium are very important semiconductor and so they are very important for elektronics.

PNICTOGENS

The pnictogens are the fifth group of the periodic table of the elements. These elements are nitrogen (N), phosphorus (P), arsenic (As), antimony (Sb) and bismuth (Bi). Nitrogen and phosphorus are nonmetals. Arsenic and antimony are metalloids and bismuth is a metal. 80% of the earth atmosphere are nitrogen, but in molecules (N_2).

CHALCOGENS

The chalcogens are the sixth group of the periodic table of the elements. These elements are oxygen (O), sulfur (S), selenium (Se), tellurium (Te) and polonium (Po). Oxygen and sulfur and nonmetals. Selenium and tellurium are metalloids and polonium is a metal.

Halogens

The halogens are the seventh group of the periodic table

of the elements. These elements are fluorine (F), chlorine (Cl), bromine (Br), iodine (I) and astatine (At). Fluorine, chlorine, bromine and iodine are nonmetals. Astatine is a metalloid.

The halogens only need one electron to reach the noble gas configuration and so they are very reactively. Fluor is the most reactively element of all elements.

Many substances (for example wood) begin to burn, because of the contact with fluorine. Fluorine is a colorless gas. The other halogens are very dangerous, because they are very reactively and very corrosive, too.

Clorine was used as a strong corrosive poison in the First World War. It is yellow-green. Bromine is a dark liquid at room temperature and like all other halogens a very corrosive substance.

NOBLE GASES

The noble gases are the eigthth group of the periodic table of the elements. These elements are helium (He), neon (Ne), argon (Ar), krypton (Kr), xenon (Xe) and radon (Rn). They are all nonmetals and like the name it says they are gases. They have got 8 outer electron (helium: 2 outer electrons) and they nearly do not react at all. They are the most noble elements.

TRANSITION METALS

We have not got mentioned the most elements of the periodic table of the elements. These are the transition metals. They metals where the other atomic shells are filled. Famous metals are for example gold (Au), silver (Ag), iron (Fe), copper (Cu), mercury (Hg) and many others. The lanthanoids (rare-earths) and the actionoids belong also to the transition metals. Famous actinoids are the both radioactive elements uranium (U) and plutonium (Pu).

Hydrogen

Elemental hydrogen is relatively rare on Earth, and is industrially produced from hydrocarbons such as methane, after which most elemental hydrogen is used "captively" (meaning locally at the production site), with the largest markets about equally divided between fossil fuel upgrading (e.g., hydrocracking) and in ammonia production (mostly for the fertilizer market). Hydrogen may be produced from water using the process of electrolysis, but this process is presently significantly more expensive commercially than hydrogen production from natural gas.

The most common naturally occurring isotope of hydrogen, known as protium, has a single proton and no neutrons. In ionic compounds it can take on either a positive charge (becoming a cation composed of a bare proton) or a negative charge (becoming an anion known as a hydride). Hydrogen can form compounds with most elements and is present in water and most organic compounds. It plays a particularly important role in acid-base chemistry, in which many reactions involve the exchange of protons between soluble molecules.

As the only neutral atom for which the Schrödinger equation can be solved analytically, study of the energetics and bonding of the hydrogen atom has played a key role in the development of quantum mechanics.

Nomenclature

Hydrogen, Latin: 'hydrogenium', is from Ancient Greek (*hydor*): "water" and (*genes*): "forming". Ancient Greek (*geinomai*): "to beget or sire")

The word "hydrogen" has several different meanings:

- The *name of an element.*
- An *atom,* sometimes called "H dot", that is abundant in space but essentially absent on Earth, because it dimerizes.
- A *diatomic molecule* that occurs naturally in trace amounts in the Earth's atmosphere; chemists increasingly refer to H_2 as *dihydrogen,*or *hydrogen molecule,* to distinguish this molecule from atomic hydrogen and hydrogen found in other compounds.
- The atomic *constituent* within all organic compounds, water, and many other chemical compounds.

The *elemental* forms of hydrogen should not be confused with hydrogen as it appears in chemical compounds.

Discovery of H_2

Hydrogen gas, H_2, was first artificially produced and formally described by T. Von Hohenheim via the mixing of metals with strong acids. He was unaware that the flammable gas produced by this chemical reaction was a new chemical element.

Robert Boyle rediscovered and described the reaction between iron filings and dilute acids, which results in the production of hydrogen gas. Henry Cavendish was the first to recognize hydrogen gas as a discrete substance, by identifying the gas from a metal-acid reaction as "inflammable air" and further finding that the gas produces water when burned.

Cavendish had stumbled on hydrogen when experimenting with acids and mercury. Although he wrongly assumed that hydrogen was a liberated component of the mercury rather than the acid, he was still able to accurately describe several key properties of hydrogen. He is usually given credit for its discovery as an element. Antoine Lavoisier gave the element the name of hydrogen when he (with Laplace) reproduced Cavendish's finding that water is produced when hydrogen is burned. Lavoisier's name for the gas won out.

Natural Occurrence

A giant region of ionized hydrogen in the Triangulum Galaxy.Hydrogen is the most abundant element in the universe, making up 75% of normal matter by mass and over 90% by number of atoms. This element is found in great abundance in stars and gas giant planets. Molecular clouds of H_2 are associated with star formation. Hydrogen plays a vital role in powering stars through proton-proton reaction nuclear fusion.

Throughout the universe, hydrogen is mostly found in the atomic and plasma states whose properties are quite different from molecular hydrogen. As a plasma, hydrogen's electron and proton are not bound together, resulting in very high electrical conductivity and high emissivity (producing the light from the sun and other stars).

The charged particles are highly influenced by magnetic and electric fields. For example, in the solar wind they interact with the Earth's magnetosphere giving rise to Birkeland currents and the aurora.

Hydrogen is found in the neutral atomic state in the Interstellar medium. The large amount of neutral hydrogen found in the damped Lyman-alpha systems is thought to dominate the cosmological baryonic density of the Universe up to redshift z = 4.

Electron Energy Levels

Depiction of a hydrogen atom showing the diameter as about twice the Bohr model radius.(Image not to scale)

The ground state energy level of the electron in a hydrogen atom is 13.6 eV, which is equivalent to an ultraviolet photon of roughly 92 nm.

The energy levels of hydrogen can be calculated fairly accurately using the Bohr model of the atom, which conceptualizes the electron as "orbiting" the proton in analogy to the Earth's orbit of the sun. However, the electromagnetic force attracts electrons and protons to one another, while planets and celestial objects are attracted to each other by gravity. Because of the discretization of angular momentum postulated in early quantum mechanics by Bohr, the electron in the Bohr model can only occupy certain allowed distances from the proton, and therefore only certain allowed energies.

A more accurate description of the hydrogen atom comes from a purely quantum mechanical treatment that uses the Schrödinger equation or the equivalent Feynman path integral formulation to calculate the probability density of the electron around the proton.

Isotopes

Protium, the most common isotope of hydrogen, has one

proton and one electron. Unique among all stable isotopes, it has no neutrons.

Hydrogen has three naturally occurring isotopes, denoted ^{1}H, ^{2}H, and ^{3}H. Other, highly unstable nuclei (^{4}H to ^{7}H) have been synthesized in the laboratory but not observed in nature.

- ^{1}H is the most common hydrogen isotope with an abundance of more than 99.98%. Because the nucleus of this isotope consists of only a single proton, it is given the descriptive but rarely used formal name *protium*.
- ^{2}H, the other stable hydrogen isotope, is known as *deuterium* and contains one proton and one neutron in its nucleus. Deuterium comprises 0.0026 – 0.0184% (by mole-fraction or atom-fraction) of hydrogen samples on Earth, with the lower number tending to be found in samples of hydrogen gas and the higher enrichments (0.015% or 150 ppm) typical of ocean water. Deuterium is not radioactive, and does not represent a significant toxicity hazard. Water enriched in molecules that include deuterium instead of normal hydrogen is called heavy water. Deuterium and its compoun126ds are used as a non-radioactive label in chemical experiments and in solvents for ^{1}H-NMR spectroscopy. Heavy water is used as a neutron moderator and coolant for nuclear reactors. Deuterium is also a potential fuel for commercial nuclear fusion.
- ^{3}H is known as *tritium* and contains one proton and two neutrons in its nucleus. It is radioactive, decaying into Helium-3 through beta decay with a half-life of 12.32 years. Small amounts of tritium occur naturally because of the interaction of cosmic rays with atmospheric gases; tritium has also been released during nuclear weapons tests. It is used in nuclear fusion reactions, as a tracer in isotope geochemistry, and specialized in self-powered lighting devices. Tritium was once routinely used in chemical and biological labeling experiments as a radiolabel.

Hydrogen is the only element that has different names for its isotopes in common use today. (During the early study of radioactivity, various heavy radioactive isotopes were given names, but such names are no longer used). The symbols D and T (instead of ^2H and ^3H) are sometimes used for deuterium and tritium, but the corresponding symbol P is already in use for phosphorus and thus is not available for protium. IUPAC states that while this use is common it is not preferred.

COMPOUNDS

Covalent and Organic Compounds

While H_2 is not very reactive under standard conditions, it does form compounds with most elements. Millions of hydrocarbons are known, but they are not formed by the direct reaction of elementary hydrogen and carbon (although synthesis gas production followed by the Fischer-Tropsch process to make hydrocarbons comes close to being an exception, as this begins with coal and the elemental hydrogen is generated in situ). Hydrogen can form compounds with elements that are more electronegative, such as halogens (e.g., F, Cl, Br, I) and chalcogens (O, S, Se); in these compounds hydrogen takes on a partial positive charge.

When bonded to fluorine, oxygen, or nitrogen, hydrogen can participate in a form of strong noncovalent bonding called hydrogen bonding, which is critical to the stability of many biological molecules. Hydrogen also forms compounds with less electronegative elements, such as the metals and metalloids, in which it takes on a partial negative charge. These compounds are often known as hydrides. Hydrogen forms a vast array of compounds with carbon. Because of their general association with living things, these compounds came to be called organic compounds; the study of their properties is known as organic chemistry and their study in the context of living organisms is known as biochemistry.

Hydrides

Compounds of hydrogen are often called hydrides, a term

that is used fairly loosely. To chemists, the term "hydride" usually implies that the H atom has acquired a negative or anionic character, denoted H". The existence of the hydride anion, suggested by G.N.

Lewis for group I and II salt-like hydrides, was demonstrated by Moers with the electrolysis of molten lithium hydride (LiH), that produced a stoichiometric quantity of hydrogen at the anode. For hydrides other than group I and II metals, the term is quite misleading, considering the low electronegativity of hydrogen. An exception in group II hydrides is BeH_2, which is polymeric. In lithium aluminum hydride, the AlH_4" anion carries hydridic centers firmly attached to the Al(III).

Although hydrides can be formed with almost all main-group elements, the number and combination of possible compounds varies widely; for example, there are over 100 binary borane hydrides known, but only one binary aluminum hydride.Binary indium hydride has not yet been identified, although larger complexes exist.

"Protons" and Acids

Oxidation of H_2 formally gives the proton, H^+. This species is central to discussion of acids, though the term proton is used loosely to refer to positively charged or cationic hydrogen, denoted H^+. A bare proton H^+ cannot exist in solution because of its strong tendency to attach itself to atoms or molecules with electrons.

To avoid the convenient fiction of the naked "solvated proton" in solution, acidic aqueous solutions are sometimes considered to contain the hydronium ion (H_3O^+) organized into clusters to form $H_9O_4^+$. Other oxonium ions are found when wat^r is in solution with other solvents. Although exotic on earth, one of the most common ions in the universe is the H_3^+ ion, known as protonated molecular hydrogen or the triatomic hydrogen cation.

Biological Syntheses

H_2 is a product of some types of anaerobic metabolism

and is produced by several microorganisms, usually via reactions catalyzed by iron- or nickel-containing enzymes called hydrogenases.

These enzymes catalyze the reversible redox reaction between H_2 and its component two protons and two electrons. Evolution of hydrogen gas occurs in the transfer of reducing equivalents produced during pyruvate fermentation to water.

Water splitting, in which water is decomposed into its component protons, electrons, and oxygen, occurs in the light reactions in all photosynthetic organisms.

Some such organisms: Including the alga *Chlamydomonas reinhardtii* and cyanobacteria — have evolved a second step in the dark reactions in which protons and electrons are reduced to form H_2 gas by specialized hydrogenases in the chloroplast.Efforts have been undertaken to genetically modify cyanobacterial hydro-genases to efficiently synthesize H_2 gas even in the presence of oxygen.

Hydrogen as an Energy Carrier

Hydrogen is not an energy source, except in the hypothetical context of commercial nuclear fusion power plants using deuterium or tritium, a technology presently far from development.

The sun's energy comes from nuclear fusion of hydrogen but this process is difficult to achieve on earth. Elemental hydrogen from solar, biological, or electrical sources costs more in energy to make than is obtained by burning it.

Hydrogen may be obtained from fossil sources (such as methane) for less energy than required to make it, but these sources are unsustainable, and are also themselves direct energy sources (and are rightly regarded as the basic source of the energy in the hydrogen obtained from them).

Molecular hydrogen has been widely discussed in the context of energy, as a possible carrier of energy on an economy-wide scale.

A theoretical advantage of using H_2 as an energy carrier is the localization and concentration of environmentally unwelcome aspects of hydrogen manufacture from fossil fuel energy sources.

GROUP 1 (ALKALI METALS)

Some atomic and physical properties of the Group 1 elements - lithium, sodium, potassium, rubidium and caesium. You will find separate sections below covering the trends in atomic radius, first ionisation energy, electronegativity, melting and boiling points, and density.The same ideas tend to recur throughout the atomic properties.

First Ionisation Energy

First ionisation energy is the energy needed to remove the most loosely held electron from each of one mole of gaseous atoms to make one mole of singly charged gaseous ions - in other words, for 1 mole of this process:

$$X_{(g)} \longrightarrow X^{+}_{(g)} + e^{-}$$

First Ionisation Energy of the Group 1 elements

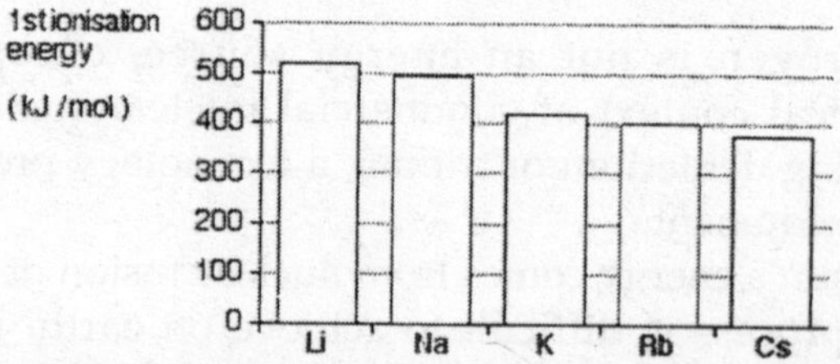

First ionisation energy falls as you go down the group.Ionisation energy is governed by:

- The charge on the nucleus,
- The amount of screening by the inner electrons,
- The distance between the outer electrons and the nucleus.

ELECTRONEGATIVITY

Electronegativity is a measure of the tendency of an atom to attract a bonding pair of electrons. It is usually measured on the Pauling scale, on which the most electronegative element (fluorine) is given an electronegativity of 4.0. All of these elements have a very low electronegativity. Remember that the most electronegative element, fluorine, has an electronegativity of 4.0.Notice that electronegativity falls as you go down the Group.

Electronegativities of the Group 1 elements

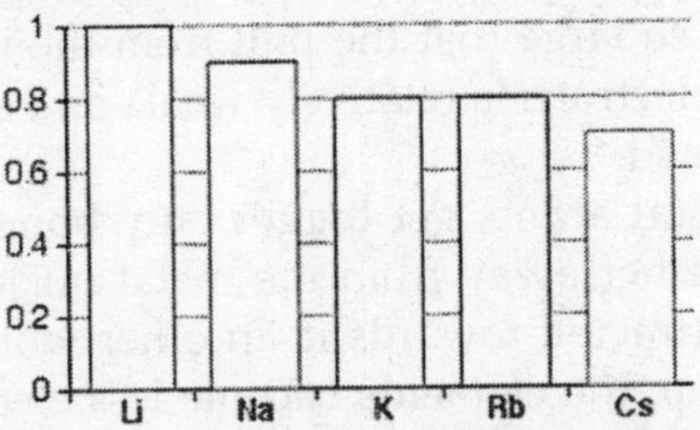

The atoms become less and less good at attracting bonding pairs of electrons.Imagine a bond between a sodium atom and a chlorine atom.

Think of it to start with as a covalent bond - a pair of shared electrons. The electron pair will be dragged towards the chlorine because there is a much greater net pull from the chlorine nucleus than from the sodium one.

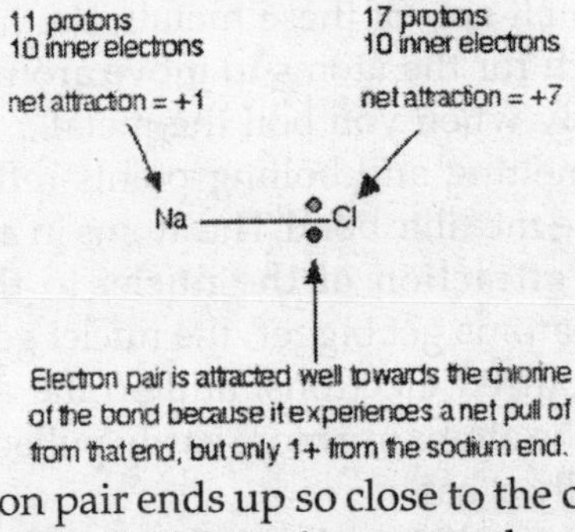

The electron pair ends up so close to the chlorine that there is essentially a transfer of an electron to the chlorine - ions are formed.

The large pull from the chlorine nucleus is why chlorine is much more electronegative than sodium is.Now compare this with the lithium-chlorine bond.

The net pull from each end of the bond is the same as before, but you have to remember that the lithium atom is smaller than a sodium atom. That means that the electron pair is going to be closer to the net 1+ charge from the lithium end, and so more strongly attracted to it. In some lithium compounds there is often a degree of covalent bonding that isn't there in the rest of the Group.

Lithium iodide, for example, will dissolve in organic solvents - a typical property of covalent compounds. The iodine atom is so large that the pull from the iodine nucleus on the pair of electrons is relatively weak, and so a fully ionic bond isn't formed.

As the metal atoms get bigger, any bonding pair gets further and further away from the metal nucleus, and so is less strongly attracted towards it. In other words, as you go down the Group, the elements become less electronegative.

With the exception of some lithium compounds, these elements all form compounds which we consider as being fully ionic. They are so weakly electronegative that we assume that the electron pair is pulled so far away towards the chlorine (or whatever) that ions are formed.

MELTING AND BOILING POINTS

When you melt any of these metals, the metallic bond is weakened enough for the atoms to move around, and is then broken completely when you boil the metal.

The fall in melting and boiling points reflects the fall in the strength of the metallic bond.The atoms in a metal are held together by the attraction of the nuclei to the delocalised electrons. As the atoms get bigger, the nuclei get further away from these delocalised electrons, and so the attractions fall. That means that the atoms are more easily pulled apart to make a liquid and finally a gas.

In the same way that we have already discussed, each of these atoms has a net pull from the nuclei of 1+. The increased charge on the nucleus as you go down the Group is offset by additional levels of screening electrons. All that matters is the distance between the nucleus and the bonding electrons.

Notice that these are all light metals - and that the first three in the Group are less dense than water (less than 1 g cm^{-}

3). That means that the first three will float on water, while the other two sink.

Density

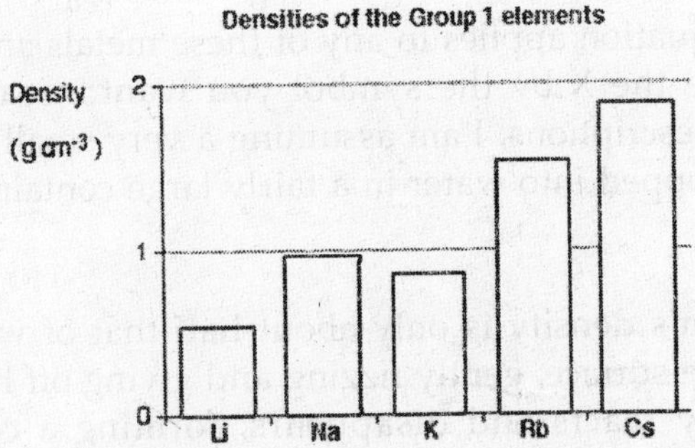

The density tends to increase as you go down the Group (apart from the fluctuation at potassium).It is quite difficult to come up with a simple explanation for this, because the density depends on two factors, both of which are changing as you go down the Group.

All of these metals have their atoms packed in the same way, so all you have to consider is how many atoms you can pack in a given volume, and what the mass of the individual atoms is. How many you can pack depends, of course, on their volume - and their volume, in turn, depends on their atomic radius.As you go down the Group, the atomic radius increases, and so the volume of the atoms increases as well. That means that you can't pack as many sodium atoms into a given volume as you can lithium atoms.However, as you go down the Group, the mass of the atoms increases.

That means that a particular number of sodium atoms will weigh more than the same number of lithium atoms.So 1 cm^3 of sodium will contain fewer atoms than the same volume of lithium, but each atom will weigh more.

REACTIONS OF GROUP 1 ELEMENTS WITH WATER

The reactions of the Group 1 elements - lithium, sodium, potassium, rubidium and caesium - with water. It uses these reactions to explore the trend in reactivity in Group 1.All of these metals react vigorously or even explosively with cold water. In each case, a solution of the metal hydroxide is produced together with hydrogen gas.

$$2X_{(s)} + 2H_2O_{(l)} \longrightarrow 2XOH_{(aq)} + H_{2(g)}$$

This equation applies to any of these metals and water - just replace the X by the symbol you want.In each of the following descriptions, I am assuming a very small bit of the metal is dropped into water in a fairly large container.

Lithium

Lithium's density is only about half that of water so it floats on the surface, gently fizzing and giving off hydrogen. It gradually reacts and disappears, forming a colourless solution of lithium hydroxide. The reaction generates heat too slowly and lithium's melting point is too high for it to melt.

Sodium

Sodium also floats on the surface, but enough heat is given off to melt the sodium (sodium has a lower melting point than lithium and the reaction produces heat faster) and it melts almost at once to form a small silvery ball that dashes around the surface. A white trail of sodium hydroxide is seen in the water under the sodium, but this soon dissolves to give a colourless solution of sodium hydroxide.

The sodium moves because it is pushed around by the hydrogen which is given off during the reaction. If the sodium becomes trapped on the side of the container, the hydrogen may catch fire to burn with an orange flame. The colour is due to contamination of the normally blue hydrogen flame with sodium compounds.

Potassium

Potassium behaves rather like sodium except that the reaction is faster and enough heat is given off to set light to the hydrogen. This time the normal hydrogen flame is contaminated by potassium compounds and so is coloured lilac (a faintly bluish pink).

Rubidium

Rubidium is denser than water and so sinks. It reacts violently and immediately, with everything spitting out of the container again. Rubidium hydroxide solution and hydrogen are formed.

Caesium

Caesium explodes on contact with water, quite possibly shattering the container. Caesium hydroxide and hydrogen are formed. The Group 1 metals become more reactive towards water as you go down the Group.

The overall enthalpy changes: The table gives estimates of the enthalpy change for each of the elements undergoing the reaction:

$$X_{(s)} + H_2O_{(l)} \longrightarrow XOH_{(aq)} + {}^1/_2\, H_{2(g)}$$

enthalpy change (kJ/ mol)

Li	–222
Na	–184
K	–196
Rb	–195
Cs	–203

When these reactions happen, the differences between them lie entirely in what is happening to the metal atoms present. In each case, you start with metal atoms in a solid and end up with metal ions in solution.

Overall, what happens to the metal is this:

$$X_{(s)} \longrightarrow X^+_{(aq)} + e^-$$

You can calculate the overall enthalpy change for this process by using Hess's Law and breaking it up into several steps that we know the enthalpy changes for.First, you would need to supply *atomisation energy* to give gaseous atoms of the metal.

$$X_{(s)} \longrightarrow X_{(g)}$$

Then ionise the metal by supplying its first *ionisation energy*.

$$X_{(g)} \longrightarrow X^{+}_{(g)} + e^{-}$$

And finally, you would get *hydration enthalpy* released when the gaseous ion comes into contact with water.

$$X^{+}_{(g)} + (aq) \longrightarrow X^{+}_{(aq)}$$

If we put values for all these steps into a table, they look like this (all values in kJ/ mol): at. Energy 1st IE hydr.

Enthalpy total:

Li	+161	+519	–519	+161
Na	+109	+494	–406	+197
K	+90	+418	–322	+186
Rb	+86	+402	–301	+187
Cs	+79	+376	–276	+179

The atomisation energy is a measure of the strength of the metallic bond in each element. This is falling as the atom gets bigger and the metallic bond is getting longer. The delocalized electrons are further from the attraction of the nuclei in the bigger atoms.

The first ionisation energy is falling because the electron being removed is getting more distant from the nucleus.The extra protons in the nucleus are screened by additional layers of electrons.The hydration enthalpy is a measure of the attraction between the metal ions and lone pairs on water molecules. As the ions get bigger, the water molecules are further from the attraction of the nucleus.

REACTIONS WITH AIR OR OXYGEN

These are all very reactive metals and have to be stored out of contact with air to prevent their oxidation. Reactivity increases as you go down the Group.Lithium, sodium and potassium are stored in oil. (Lithium in fact floats on the oil, but there will be enough oil coating it to give it some protection. It is, anyway, less reactive than the rest of the Group.)Rubidium and caesium are normally stored in sealed glass tubes to prevent air getting at them.

They are stored either in a vacuum or in an inert atmosphere of, say, argon. The tubes are broken open when the metal is used.Depending on how far down the Group you are, different kinds of oxide are formed when the metals burn. Reaction with oxygen is just a more dramatic version of the reaction with air.Lithium is unique in the Group because it also reacts with the nitrogen in the air to form lithium nitride.

Lithium

Lithium burns with a strongly red-tinged flame if heated in air. It reacts with oxygen in the air to give white lithium oxide. With pure oxygen, the flame would simply be more intense.

$$4Li + O_2 \longrightarrow 2Li_2O$$

For the record, it also reacts with the nitrogen in the air to give lithium nitride. Lithium is the only element in this Group to form a nitride in this way.

$$6Li + N_2 \longrightarrow 2Li_3N$$

Sodium

Small pieces of sodium burn in air with often little more than an orange glow. Using larger amounts of sodium or burning it in oxygen gives a strong orange flame. You get a white solid mixture of sodium oxide and sodium peroxide.The equation for the formation of the simple oxide is just like the lithium one.

$$4Na + O_2 \longrightarrow 2Na_2O$$

The peroxide equation is:

$$2Na + O_2 \longrightarrow Na_2O_2$$

Potassium: Small pieces of potassium heated in air tend to just melt and turn instantly into a mixture of potassium peroxide and potassium superoxide without any flame being seen. Larger pieces of potassium burn with a lilac flame.

The equation for the formation of the peroxide is just like the sodium one above:

$$2K + O_2 \longrightarrow K_2O_2$$

... and for the superoxide:

$$K + O_2 \longrightarrow KO_2$$

Rubidium and caesium: Both metals catch fire in air and produce superoxides, RbO_2 and CsO_2. The equations are the same as the equivalent potassium one.Both superoxides are described in most sources as being either orange or yellow.

Lithium (and to some extent sodium) form simple oxides, X_2O, which contain the common O^{2-} ion. Sodium (and to some extent potassium) form peroxides, X_2O_2, containing the more complicated O_2^{2-} ion. Potassium, rubidium and caesium form superoxides, XO_2.

The structure of the superoxide ion, O_2^-, is too difficult to discuss at this level, needing a good knowledge of molecular orbital theory to make sense of it. The more complicated ions aren't stable in the presence of a small positive ion. Consider the peroxide ion, for example.

The peroxide ion, O_2^{2-} looks like this:

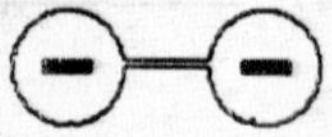

Fig. Two oxygen Atoms Each Carrying a Negative Change by a Covalent Bond.

The covalent bond between the two oxygen atoms is relatively weak. Now imagine bringing a small positive ion close to the peroxide ion. Electrons in the peroxide ion will be strongly attracted towards the positive ion. This is then well on the way to forming a simple oxide ion if the right-hand oxygen atom breaks off.

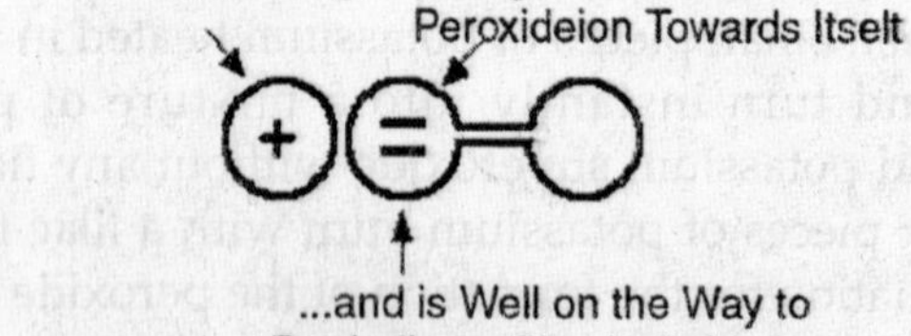

We say that the positive ion *polarises* the negative ion. This works best if the positive ion is small and highly charged - if it has a *high charge density*.Even though it only has one charge, the lithium ion at the top of the Group is so small and has such a high charge density that any peroxide ion near it falls to pieces to give an oxide and oxygen.

As you go down the Group to sodium and potassium the positive ions get bigger and they don't have so much effect on the peroxide ion. The superoxide ions are even more easily pulled apart, and these are only stable in the presence of the big ions towards the bottom of the Group.So why do any of the metals form the more complicated oxides?

It is a matter of energetics.In the presence of sufficient oxygen, they produce the compound whose formation gives out most energy. That gives the most stable compound.The amount of heat evolved per mole of rubidium in forming its various oxides is: enthalpy change (kJ/ mol of Rb)

Rb_2O	–169.5
Rb_2O_2	–236
RbO_2	–278.7

The values for the various potassium oxides show exactly the same trends. As long as you have enough oxygen, forming the peroxide releases more energy per mole of metal than forming the simple oxide. Forming the superoxide releases even more.Forming the more complicated oxides from the metals releases more energy and makes the system more energetically stable. BUT... this only works for the metals in the lower half of the Group where the metal ions are big and have a low charge density.At the top of the Group, the small ions with a higher charge density tend to polarise the more complicated oxide ions to the point of destruction.

Reaction with Water

These are simple basic oxides, reacting with water to give the metal hydroxide.For example, lithium oxide reacts with water to give a colourless solution of lithium hydroxide.

$$X_2O + H_2O \longrightarrow 2XOH$$

Reaction with Dilute Acids

These simple oxides all react with an acid to give a salt

and water. For example, sodium oxide will react with dilute hydrochloric acid to give colourless sodium chloride solution and water.

$$X_2O + 2HCl \longrightarrow 2XCl + H_2O$$

Reaction with water: If the reaction is done ice cold (and the temperature controlled so that it doesn't rise even though these reactions are strongly exothermic), a solution of the metal hydroxide and hydrogen peroxide is formed.

$$X_2O_2 + 2H_2O \longrightarrow 2XOH + H_2O_2$$

If the temperature increases (as it inevitably will unless the peroxide is added to water very, very, very slowly!), the hydrogen peroxide produced decomposes into water and oxygen. The reaction can be very violent overall.

Reaction with Dilute Acids

These reactions are even more exothermic than the ones with water. A solution containing a salt and hydrogen peroxide is formed. The hydrogen peroxide will decompose to give water and oxygen if the temperature rises - again, it is almost impossible to avoid this. Another potentially violent reaction!

$$X_2O_2 + 2HCl \longrightarrow 2XCl + H_2O_2$$

Reaction with Water

This time, a solution of the metal hydroxide and hydrogen peroxide is formed, but oxygen gas is given off as well. Once again, these are strongly exothermic reactions and the heat produced will inevitably decompose the hydrogen peroxide to water and more oxygen. Again violent!

$$2XO_2 + 2H_2O \longrightarrow 2XOH + H_2O_2 + O_2$$

Reaction with Dilute Acids

Again, these reactions are even more exothermic than the ones with water. A solution containing a salt and hydrogen peroxide is formed together with oxygen gas. The hydrogen peroxide will again decompose to give water and oxygen as the temperature rises. Violent!

$$2XO_2 + 2HCl \longrightarrow 2XCl + H_2O_2 + O_2$$

The Reactions of the Elements with Chlorine

This is included because of the similarity in appearance between the reactions of the Group 1 metals with chlorine and with oxygen.Sodium, for example, burns with an intense orange flame in chlorine in exactly the same way that it does in pure oxygen.

The rest also behave the same in both gases.In each case, there is a white solid residue which is the simple chloride, XCl. There is nothing in any way complicated about these reactions!

$$2X + Cl_2 \longrightarrow 2XCl$$

Some compounds of the Group 1 elements (lithium, sodium, potassium, rubidium and caesium). You will find some information about the nitrates, carbonates, hydrogencarbonates and hydrides of the metals.We will first look at what happens to some of the compounds on heating, and then their solubility.

Heating the Nitrates

Most nitrates tend to decompose on heating to give the metal oxide, brown fumes of nitrogen dioxide, and oxygen.

For example, a typical *Group 2 nitrate* like magnesium nitrate decomposes like this:

$$2Mg(NO_3)_{2(s)} \longrightarrow 2MgO_{(s)} + 4NO_{2(g)} + O_{2(g)}$$

In *Group 1*, lithium nitrate behaves in the same way - producing lithium oxide, nitrogen dioxide and oxygen.

$$4LiNO_{3(s)} \longrightarrow 2Li_2O_{(s)} + 4NO_{2(g)} + O_{2(g)}$$

The rest of the Group, however, don't decompose so completely (at least not at Bunsen temperatures) - producing the metal nitrite and oxygen, but no nitrogen dioxide.

$$2XNO_{3(s)} \longrightarrow 2XNO_{2(s)} + O_{2(g)}$$

All the nitrates from sodium to caesium decomposes in this same way, the only difference being how hot they have to be to undergo the reaction. As you go down the Group, the decomposition gets more difficult, and you have to use higher temperatures.

Heating the Carbonates

Most carbonates tend to decompose on heating to give the metal oxide and carbon dioxde.For example, a typical *Group 2 carbonate* like calcium carbonate decomposes like this:

$$CaCO_{3(s)} \longrightarrow CaO_{(s)} + CO_{2(g)}$$

In *Group 1*, lithium carbonate behaves in the same way - producing lithium oxide and carbon dioxide.

$$Li_2CO_{3(s)} \longrightarrow Li_2O_{(s)} + CO_{2(g)}$$

The rest of the Group 1 carbonates don't decompose at Bunsen temperatures, although at higher temperatures they will. The decomposition temperatures again increase as you go down the Group.

The Thermal Stability of the Hydrogencarbonates

The *Group* 2 hydrogencarbonates like calcium hydrogen-carbonate are so unstable to heat that they only exist in solution. Any attempt to get them out of solution causes them to decompose to give the carbonate, carbon dioxide and water.

$$Ca(HCO_3)_{2(aq)} \longrightarrow CaCO_{3(s)} + CO_{2(g)} + H_2O_{(l)}.$$

By contrast, the *Group 1* hydrogencarbonates are stable enough to exist as solids, although they do decompose easily on heating.

For example, for sodium hydrogencarbonate:

$$2NaHCO_{3(s)} \longrightarrow Na_2CO_{3(s)} + CO_{2(g)} + H_2O_{(l)}$$

Detailed explanations are given for the carbonates because the diagrams are easier to draw. Exactly the same arguments apply to the nitrates or hydrogencarbonates.There are two ways of explaining the increase in thermal stability as you go down the Group.

The hard way is in terms of the energetics of the process; the simple way is to look at the polarising ability of the positive ions.A small positive ion has a lot of charge packed into a small volume of space - especially if it has more than one positive charge. It has a high *charge density* and will have a marked distorting effect on any negative ions which happen to be near it.A bigger positive ion has the same charge spread over a larger volume of space. Its charge density will be lower, and it will cause less distortion to nearby negative ions.

Carbonate ion

If you worked out the structure of a carbonate ion using "dots-and-crosses" or some similar method, you would probably come up with:

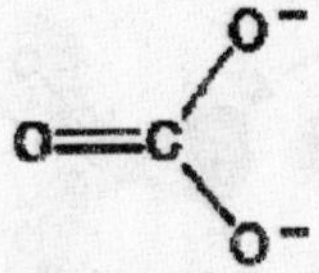

This shows two single carbon-oxygen bonds and one double one, with two of the oxygens each carrying a negative charge. Unfortunately, in real carbonate ions all the bonds are identical, and the charges are spread out over the whole ion - although concentrated on the oxygen atoms. We say that the charges are *delocalised*.

This is a rather more complicated version of the bonding you might have come across in benzene or in ions like ethanoate. For the purposes of this topic, you don't need to understand how this bonding has come about. The next diagram shows the delocalised electrons. The shading is intended to show that there is a greater chance of finding them around the oxygen atoms than near the carbon.

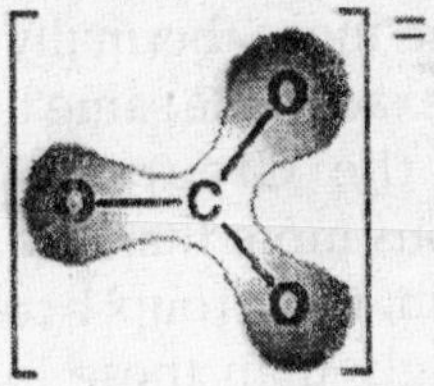

Polarising the Carbonate Ion

Now imagine what happens when this ion is placed next to a positive ion. The positive ion attracts the delocalised electrons in the carbonate ion towards itself. The carbonate ion becomes polarised. The diagram shows what happens with an ion from Group 2, carrying two positive charges. If this is heated, the carbon dioxide breaks free to leave the metal oxide.How much you need to heat the carbonate before that happens depends on how polarised the ion was. If it is highly polarised, you need less heat than if it is only slightly polarised.If the positive ion only had one positive charge, the polarising effect would be less.

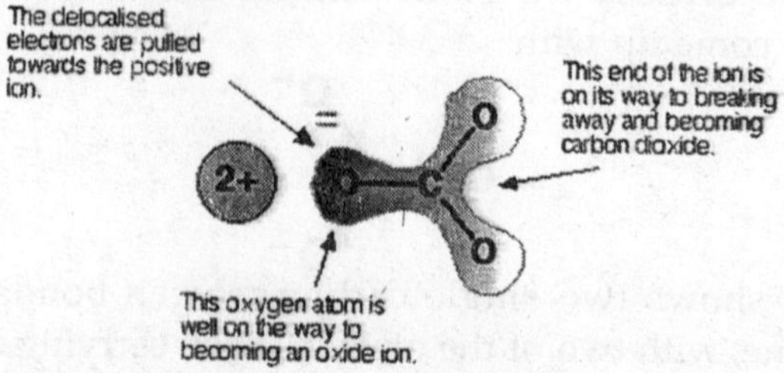

That is why the: Group 1 compounds are more thermally stable than those in Group 2. You have to heat the Group 1 compound more because the carbonate ions are less polarised by singly charged positive ions.

The smaller the positive ion is, the higher the charge density, and the greater effect it will have on the carbonate ion. As the positive ions get bigger as you go down the Group, they have less effect on the carbonate ions near them. To compensate for that, you have to heat the compound more in order to persuade the carbon dioxide to break free and leave the metal oxide.In other words, as you go down the Group, the carbonates become more thermally stable.

The argument is exactly the same here. The small positive ions at the top of the Group polarise the nitrate or hydrogencarbonate ions more than the larger positive ions at the bottom. And, again, the Group 1 compounds will need to be heated more strongly than those in Group 2 because the Group 1 ions are less polarising.

ELECTROLYSIS

On heating, most of these hydrides decompose back into the metal and hydrogen before they melt. It is, however, possible to melt lithium hydride and to electrolyse the melt. The metal is released at the cathode as you would expect. Hydrogen is given off at the anode (the positive electrode) and this is evidence for the presence of the negative hydride ion in lithium hydride.

The anode equation is:

$$2H^- \longrightarrow H_2 + 2e^-$$

The other Group 1 hydrides can be electrolysed in solution in various molten mixtures such as a mixture of lithium chloride and potassium chloride. Mixtures such as these melt at lower temperatures than the pure chlorides.

Reaction with Water

These hydrides react violently with water releasing hydrogen gas and producing the metal hydroxide.For example, sodium hydride reacts with water to produce a solution of sodium hydroxide and hydrogen gas.

$$NaH + H_2O \longrightarrow NaOH + H_2$$

FLAME TESTS

A flame test for a range of metal ions, and briefly describes how the flame colour arises.Flame tests are used to identify the presence of a relatively small number of metal ions in a compound. Not all metal ions give flame colours.For Group 1 compounds, flame tests are usually by far the easiest way of identifying which metal you have got. For other metals, there are usually other easy methods which are more reliable - but the flame test can give a useful hint as to where to look.

Carrying out a Flame Test

Clean a platinum or nichrome (a nickel-chromium alloy) wire by dipping it into concentrated hydrochloric acid and then holding it in a hot Bunsen flame. Repeat this until the wire doesn't produce any colour in the flame.When the wire is clean, moisten it again with some of the acid and then dip it into a small amount of the solid you are testing so that some sticks to the wire. Place the wire back in the flame again.If the flame colour is weak, it is often worthwhile to dip the wire back in the acid again and put it back into the flame as if you were cleaning it.

The Colours

The colours in the table are just a guide. Almost everybody sees and describes colours differently. for example, used the word "red" several times to describe colours which can be quite different from each other.

Flame	Colour
Li	red
Na	strong persistent orange
K	lilac (pink)
Rb	red (reddish-violet)
Cs	blue violet
Ca	orange-red
Sr	red
Ba	pale green
Cu	blue-green (often with white flashes)
Pb	greyish-white

The Origin of Flame Colours

Flame colours are produced from the movement of the electrons in the metal ions present in the compounds. For example, a sodium ion in an unexcited state has the structure $1s^22s^22p^6$.When you heat it, the electrons gain energy and can jump into any of the empty orbitals at higher levels - for example, into the 7s or 6p or 4d or whatever, depending on how much energy a particular electron happens to absorb from the flame.Because the electrons are now at a higher and more energetically unstable level, they tend to fall back down to where they were before - but not necessarily all in one go.

An electron which had been excited from the 2p level to an orbital in the 7 level, for example, might jump back to the 2p level in one go. That would release a certain amount of energy which would be seen as light of a particular colour. However, it might jump back in two (or more) stages.

For example, first to the 5 level and then back to the 2 level. Each of these jumps involves a specific amount of energy being released as light energy, and each corresponds to a particular colour.As a result of all these jumps, a spectrum of coloured lines will be produced.

The colour you see will be a combination of all these individual colours. The exact sizes of the possible jumps in energy terms vary from one metal ion to another. That means that each different ion will have a different pattern of spectral lines, and so a different flame colour.

GROUP 2 (ALKALINE EARTH METALS)

The trends in some atomic and physical properties of the Group 2 elements - beryllium, magnesium, calcium, strontium and barium. The trends in atomic radius, first ionisation energy, electronegativity and melting point.Even if you aren't currently interested in all these things. The same ideas tend to recur throughout the atomic properties.

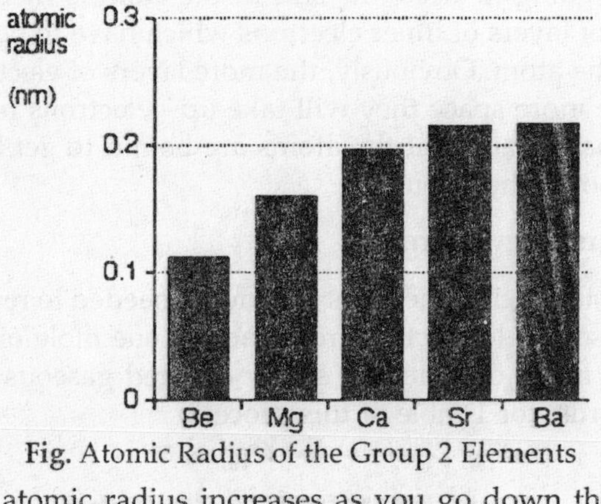

Fig. Atomic Radius of the Group 2 Elements

The atomic radius increases as you go down the Group. Notice that beryllium has a particularly small atom compared with the rest of the Group.

The radius of an atom is governed by:

- The number of layers of electrons around the nucleus
- The pull the outer electrons feel from the nucleus.

Compare beryllium and magnesium:

- Be$1s^2 2s^2$
- Mg $1s^2 2s^2 2p^6 3s^2$

In each case, the two outer electrons feel a net pull of 2+ from the nucleus. The positive charge on the nucleus is cut down by the negativeness of the inner electrons.

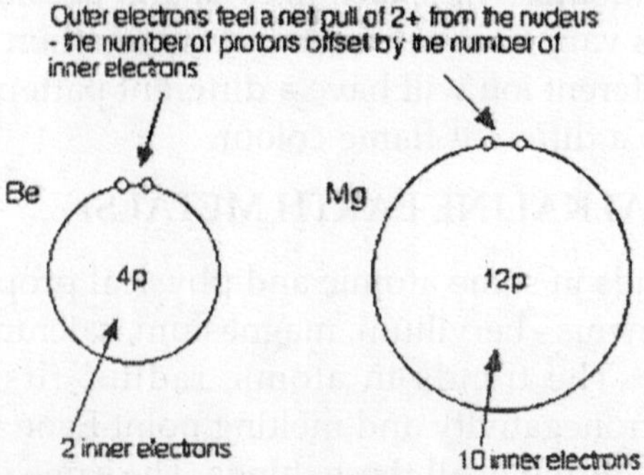

This is equally true for all the other atoms in Group 2. Work it out for calcium if you aren't convinced. The only factor which is going to affect the size of the atom is therefore the number of layers of inner electrons which have to be fitted in around the atom. Obviously, the more layers of electrons you have, the more space they will take up - electrons repel each other. That means that the atoms are bound to get bigger as you go down the Group.

First Ionisation Energy

First ionisation energy is the energy needed to remove the most loosely held electron from each of one mole of gaseous atoms to make one mole of singly charged gaseous ions - in other words, for 1 mole of this process:

$$X_{(g)} \longrightarrow X^{+}_{(g)} + e^{-}$$

First Ionisation Energy of the Group 2 elements

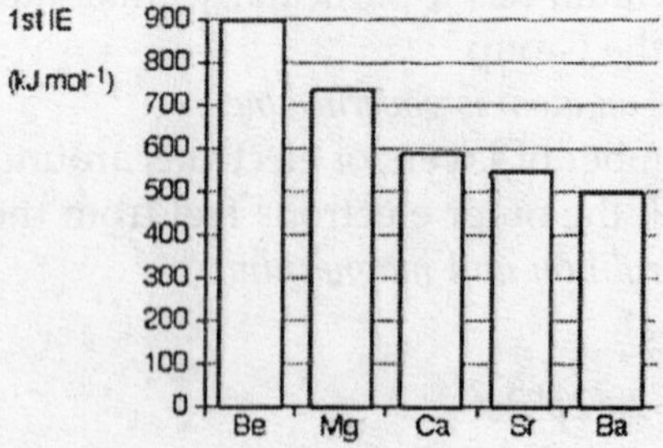

Notice that first ionisation energy falls as you go down the group.

Ionisation energy is governed by:

- The charge on the nucleus,
- The amount of screening by the inner electrons,
- The distance between the outer electrons and the nucleus.

As you go down the Group, the increase in nuclear charge is exactly offset by the increase in the number of inner electrons. Just as when we were talking about atomic radius further up, in each of the elements in this Group, the outer electrons feel a net attraction of 2+ from the centre.

However, as you go down the Group, the distance between the nucleus and the outer electrons increases and so they become easier to remove - the ionisation energy falls.

ELECTRONEGATIVITY

Electronegativity is a measure of the tendency of an atom to attract a bonding pair of electrons. It is usually measured on the Pauling scale, on which the most electronegative element (fluorine) is given an electronegativity of 4.0.

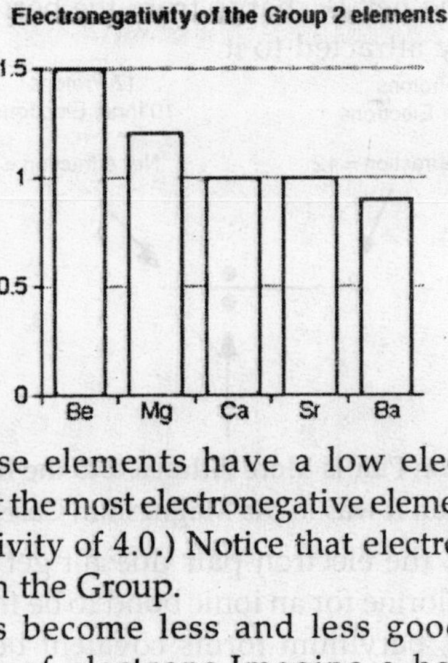

All of these elements have a low electronegativity. (Remember that the most electronegative element, fluorine, has an electronegativity of 4.0.) Notice that electronegativity falls as you go down the Group.

The atoms become less and less good at attracting bonding pairs of electrons.Imagine a bond between a magnesium atom and a chlorine atom. Think of it to start with as a covalent bond - a pair of shared electrons. The electron pair will be dragged towards the chlorine end because there is a much greater net pull from the chlorine nucleus than from the magnesium one.

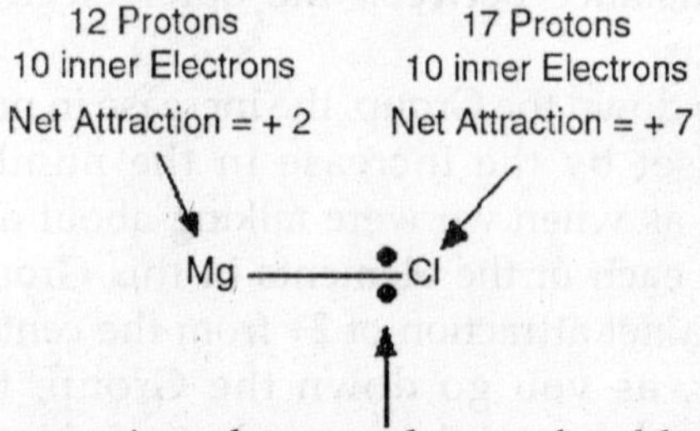

The electron pair ends up so close to the chlorine that there is essentially a transfer of an electron to the chlorine - ions are formed.The large pull from the chlorine nucleus is why chlorine is much more electronegative than magnesium is.

Now compare this with the beryllium-chlorine bond.The net pull from each end of the bond is the same as before, but you have to remember that the beryllium atom is smaller than a magnesium atom. That means that the electron pair is going to be closer to the net 2+ charge from the beryllium end, and so more strongly attracted to it.

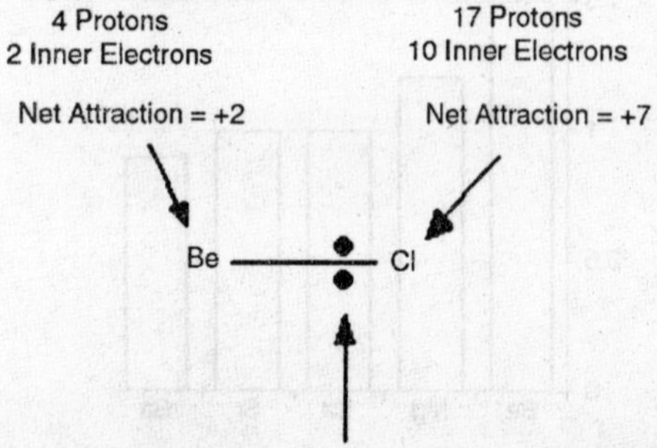

Fig. Electron Pair is More Attracted to the Beryllium than it was in the Magnesium Case.

In this case, the electron pair doesn't get attracted close enough to the chlorine for an ionic bond to be formed. Because of its small size, beryllium forms covalent bonds, not ionic ones. The attraction between the beryllium nucleus and a bonding pair is always too great for ions to be formed. As the metal atoms get bigger, any bonding pair gets further and further away from the metal nucleus, and so is less strongly attracted towards it.

In other words, as you go down the Group, the elements become less electronegative.As you go down the Group, the bonds formed between these elements and other things such as chlorine become more and more ionic. The bonding pair is increasingly attracted away from the Group 2 element towards the chlorine.

Melting Point

With the exception of magnesium the melting point falls as you go down the Group. It would be tempting to say that apart from magnesium the lower melting points as you go down the Group reflect weaker metallic bonds, but this is dangerous.

The metallic bond isn't fully broken on melting. Boiling point is usually a better guide to the strengths of the forces involved. You will sometimes come across an explanation for the low melting point of magnesium in terms of the way the atoms pack in the crystal.

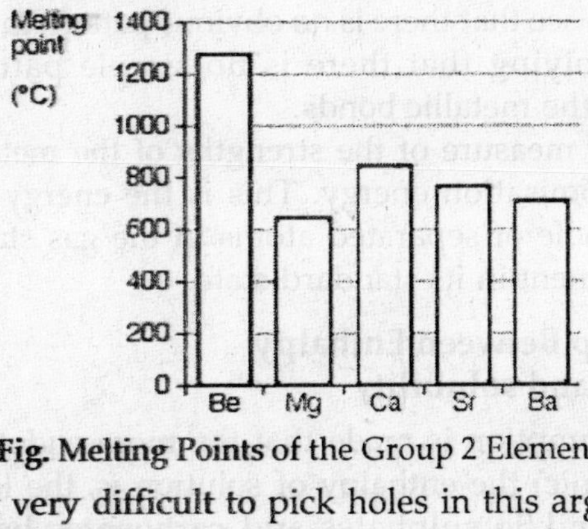

Fig. Melting Points of the Group 2 Elements

It isn't very difficult to pick holes in this argument and, in any case, the boiling point of magnesium is also surprisingly low - and packing is completely irrelevant when you are thinking about a liquid. In the case of magnesium, there must be some factor producing a relatively weak metallic bond.

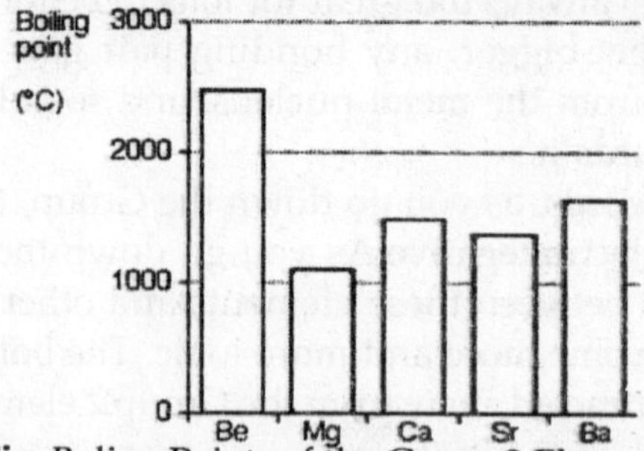

Fig. Boling Points of the Group 2 Elements

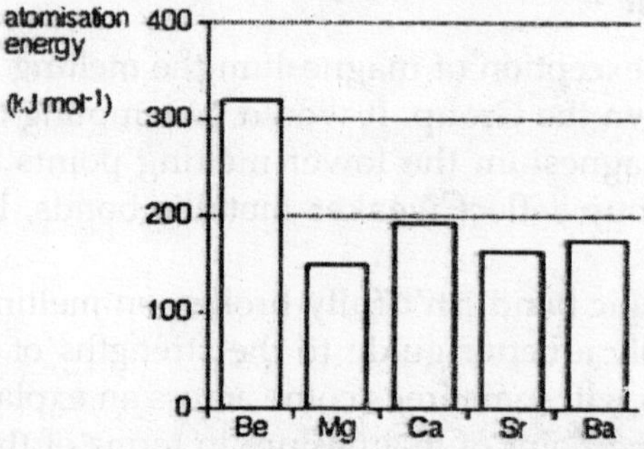

Fig. Atomisation Energy of the Group 2 Elements

You will see that there is no obvious pattern in the boiling points - implying that there is no simple pattern in the strengths of the metallic bonds.

Another measure of the strengths of the metallic bonds might be atomisation energy. This is the energy needed to produce 1 mole of separated atoms in the gas state starting from the element in its standard state.

Relationship Between Enthalpy of Solution and solubility

The assumption is made that the more endothermic (or less exothermic) the enthalpy of solution is, the less soluble the compound.So sulphates and carbonates become less soluble as you go down the Group; hydroxides become more soluble. The enthalpy of solution values for the Group 1 chlorides.

	Calculated value (kJ mol^{-1})	Data book value(kJ mol^{-1})
NaCl	+1	+3.9
KCl	+15	+17.2
RbCl	+10	+16.7
CsCl	+5	+17.9

The reasons for the discrepancies lie in the way the numbers are calculated. These are very small numbers worked out from much larger ones. Small uncertainties in those large numbers will cause large swings in the answers.The solubilities of the Group 1 chlorides (in moles of solute saturating 100 g of water at 298 K) compared with their enthalpies of solution are:

	Calculated value (kJ mol^{-1})	Solubility (mol/100 g of water)
NaCl	+3.9	0.615
KCl	+17.2	0.481
RbCl	+16.7	0.781
CsCl	+17.9	1.13

There is no obvious relationship connecting the relative movements of these solubility values with the enthalpy of solution figures. It would be quite untrue to say that the more endothermic the change, the less soluble the compound.The table above illustrates this problem.Taking the sign of enthalpy of solution at face value.

$MgCO_3$	–25.3
$CaCO_3$	–12.3
$SrCO_3$	–3.4
$BaCO_3$	+4.2

Sodium chloride and the other Group 1 chlorides dissolve despite the fact that their enthalpies of solution are positive, and yet magnesium carbonate (and most of the other Group 2 carbonates) are very sparingly soluble, but have exothermic enthalpies of solution.The Group 1 chlorides, you could argue that the figures get progressively less exothermic, and at barium carbonate become endothermic. That would seem to support the decrease in solubility as you go down the Group quite nicely.If you look at the solubility data, the trend is broken at the bottom of the Group. Barium carbonate is more soluble than strontium carbonate.

To get around the problem of many compounds dissolving freely in water despite the fact that their enthalpies of solution are endothermic you have to introduce the concept of *entropy change.* The only way of making sense of entropy without getting bogged down in some serious maths is to think of it as a measure of the amount of disorder in a system.

Entropy is given the symbol S. If a system becomes more disordered, then its entropy increases.In order to see whether a change is possible or not, you have to think about a combination of the enthalpy change and the entropy change. These can be combined mathematically to give an important term known as *free energy change.*

Enthalpy Change   Entropy Change

$$\Delta G = \Delta H - T\Delta S$$

Free Energy Change   Temperature (K)

As an approximation, for a reaction to happen, the free energy change must be negative.What happens if the enthalpy change is positive - as for example when sodium chloride dissolves in water.

As long as the entropy change is positive enough, it is possible to get a negative value for free energy change. In the sodium chloride case, you don't have to have very much increase in entropy to outweigh the small enthalpy change.

So... does the entropy increase when sodium chloride dissolve in water? Yes, it does! The sodium and chloride ions were arranged in a very tidy way in the crystal lattice - their entropy was low.When you dissolve the crystal in water, the entropy increases as the ions and water molecules become completely jumbled up - they become much more disordered than they were originally.

$$\Delta G = \Delta H - T\Delta S$$

If entrophy change is postive enough, the negative value of - TΔS will outweigh the postive value of ΔH

ΔG will then become negative, and the change is feasible.

The most obvious thing that's wrong is that it won't explain why some compounds (like magnesium carbonate, and most of the other Group 2 carbonates) don't dissolve in water even though their enthalpies of solution are mainly negative.In these cases, the entropy of the system must fall when the compounds dissolve in water - in other words, the solution in water is more ordered than the original crystal and water!

If this is negative the only way of making ΔG postive is for - TΔS to be postive

$$\Delta G = \Delta H - T\Delta S$$

That only happens if DS is negative - if the system becomes more ordered than it was before.

ΔG must be postive because the change isn't happening to any extent.

This happens because the water molecules become more ordered when the compound dissolves in them. Instead of milling around pretty much at random, they become attracted to the ions present and arranged around them.

This is particularly effective if the ions are small and highly charged - and so the effect is greatest for the positive ions at the top of the Group, and gets less as you go down.

It is also much more important in Group 2 than in Group 1 where the ions only carry one positive charge.That means that you have two entropy effects to consider.

There is the increase in disorder as the crystal lattice breaks up, but a corresponding increase in order in the water - which varies depending on the sizes and charges of the ions present.

To explain this properly, you need to think about the way lattice enthalpy changes as you go down the Group, the way that hydration enthalpies change, and the way that entropy changes. The way those changes happen will vary from one type of compound to another.

Remember that the solubility of the carbonates falls as you go down Group 2, apart from an increase as you go from strontium to barium carbonate.The general fall is because hydration enthalpies are falling faster than lattice enthalpies. Remember that where you have a big negative ion, its size dominates the inter-ionic distance and so doesn't allow the lattice enthalpy to change much.

This gives the enthalpy of solution values we've already looked at (values in kJ mol^{-1}):

$MgCO_3$	–25.3
$CaCO_3$	–12.3
$SrCO_3$	–3.4
$BaCO_3$	+4.2

But the entropy change will also be varying as you go down the Group. At the top, where you have small 2+ ions, the overall entropy change in the system must be negative - the system as a whole becomes more ordered when the compound dissolves because of the way the water molecules become organised around the positive ions. That negative entropy change is going to be enough to wipe out the effect of the exothermic enthalpy of solution.

THERMAL STABILITY

The effect of heat on the carbonates and nitrates of the Group 2 elements - beryllium, magnesium, calcium, strontium and barium. It describes and explains how the thermal stability of the compounds changes as you go down the Group. All the carbonates in this Group undergo thermal decomposition to give the metal oxide and carbon dioxide. Thermal decomposition is the term given to splitting up a compound by heating it.If "X" represents any one of the elements:

$$XCO_{3(s)} \longrightarrow XO_{(s)} + CO_{2(g)}$$

As you go down the Group, the carbonates have to be heated more strongly before they will decompose.

The carbonates become more stable to heat as you go down the Group. All the nitrates in this Group undergo thermal decomposition to give the metal oxide, nitrogen dioxide and oxygen.

Again, if "X" represents any one of the elements:

$$2X(NO_3)_{2(s)} \longrightarrow 2XO_{(s)} + 4NO_{2(g)} + O_{2(g)}$$

As you go down the Group, the nitrates also have to be heated more strongly before they will decompose.

The nitrates also become more stable to heat as you go down the Group. Both carbonates and nitrates become more thermally stable as you go down the Group. The ones lower down have to be heated more strongly than those at the top before they will decompose.A small 2+ ion has a lot of charge packed into a small volume of space. It has a high charge density and will have a marked distorting effect on any negative ions which happen to be near it.A bigger 2+ ion has the same charge spread over a larger volume of space. Its charge density will be lower, and it will cause less distortion to nearby negative ions.

Carbonate Ion

If you worked out the structure of a carbonate ion using "dots-and-crosses" or some similar method, you would probably come up with:

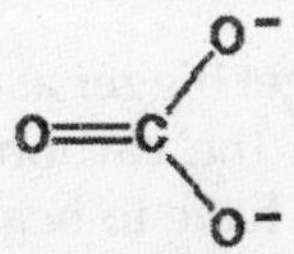

This shows two single carbon-oxygen bonds and one double one, with two of the oxygens each carrying a negative charge. Unfortunately, in real carbonate ions all the bonds are identical, and the charges are spread out over the whole ion - although concentrated on the oxygen atoms. We say that the charges are delocalised.This is a rather more complicated version of the bonding you might have come across in benzene or in ions like ethanoate.

The next diagram shows the delocalised electrons. The shading is intended to show that there is a greater chance of finding them around the oxygen atoms than near the carbon.

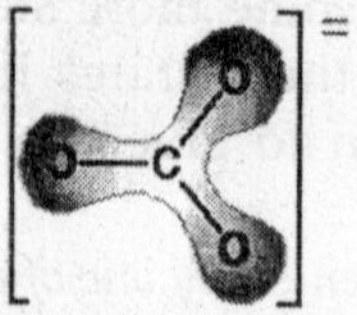

Polarising the carbonate ion: Now imagine what happens when this ion is placed next to a positive ion. The positive ion attracts the delocalised electrons in the carbonate ion towards itself. The carbonate ion becomes polarised.

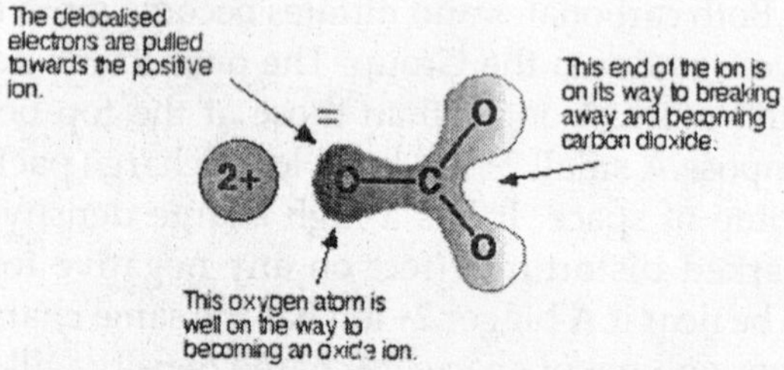

If this is heated, the carbon dioxide breaks free to leave the metal oxide.How much you need to heat the carbonate before that happens depends on how polarised the ion was. if it is highly polarised, you need less heat than if it is only slightly polarised.The smaller the positive ion is, the higher the charge density, and the greater effect it will have on the carbonate ion.

As the positive ions get bigger as you go down the Group, they have less effect on the carbonate ions near them. To compensate for that, you have to heat the compound more in order to persuade the carbon dioxide to break free and leave the metal oxide.In other words, as you go down the Group, the carbonates become more thermally stable.

The argument is exactly the same here. The small positive ions at the top of the Group polarise the nitrate ions more than the larger positive ions at the bottom. Drawing diagrams to show this happening is much more difficult because the process has interactions involving more than one nitrate ion. If you calculate the enthalpy changes for the decompostion of the various carbonates, you find that all the changes are quite strongly endothermic.

That implies that the reactions are likely to have to be heated constantly to make them happen.The calculated enthalpy changes (in kJ mol^{-1}) are given in the table. Figures to calculate the beryllium carbonate value weren't available. Remember that the reaction we are talking about is:

$$XCO_{3(s)} \longrightarrow XO_{(s)} + CO_{2(g)}$$

$MgCO_3$	+117
$CaCO_3$	+178
$SrCO_3$	+235
$BaCO_3$	+267

The reactions become more endothermic as you go down the Group. That's entirely what you would expect as the carbonates become more thermally stable.

Hydroxide ions, OH: These react with hydrogen ions from an acid to form water - and so the hydroxide reacts with acids.There isn't any equivalent to the neutral complex. Adding more hydroxide ions from a base has no effect because they haven't got anything to react with.

Chapter 6

Light and Matter

QUANTIZED ATOM

This implies that the atom is only allowed to have certain energies. We now know that the electric field associated with the light photon interacts with the electrons of the atom.

When a photon is absorbed by an atom, the energy of the photon is used to cause an electron of the atom to make a transition ("jump") to a higher energy state within the atom. The photon that was absorbed disappears. Only certain energy states are allowed to the electrons.

That is why only photons of certain frequencies (energies) can be absorbed. The photon energy is exactly the same as the energy *difference* between 2 allowed electron states.

We write this as:

$$DE = E(\text{state } 2) - E(\text{state } 1) = hn$$

This equation has been tremendously important in the development of our understanding of the detailed structures of atoms and molecules. Virtually everything we know about the atom, its internal structure, and the structures of molecules containing several atoms, is a result of studying the interaction of light with the atom or molecule. For you to have a qualitative understanding of this interaction is therefore very important.

ABSORPTION OF LIGHT

We begin this section by stating some very important ideas, which will then be illustrated in subsequent discussion.

The ideas are these:

- Light consists of oscillating electric and magnetic fields.
- Because nuclei and electrons are charged particles, their motions in atoms and molecules generate oscillating electric fields.
- An atom or molecule can absorb energy from light if the frequency of the light oscillation and the frequency of the electron or molecular "transition motion" match. Unless these frequencies match, light absorption cannot occur. The "transition motion" frequency is related to the frequencies of motion in the higher and lower energy states.
- By measuring the frequencies of light absorbed by an atom or molecule, we can determine the frequencies of the various transition motions that the atom or molecule can have. Thus we can use light absorption to probe the dynamics of atoms and molecules!

These ideas are the basis for the techniques of *spectroscopy,* the study of the interaction of light and matter. The word means "measurement of spectrum," where the spectrum of a substance is the array of frequencies of light absorbed by its atoms or molecules.

You should already be familiar with the idea of energy absorption through matching frequencies. This is exactly the way in which you transfer energy to a child on a swing. You push at a frequency that matches that of the swing.

Energy is transferred successfully, making the swing go to a higher energy state (higher amplitude), only if the push is made at the right time. If the frequency of pushing does not match the frequency of swinging, very little if any energy is transferred. In fact, the swing may actually lose energy if you try to push it while it is moving back toward you. This idea operates in the same way when light and matter interact. It is complicated only by the fact that, in contrast to the swing, molecules can undergo more than one type of motion.

Molecular Motions and Energy

There are many ways in which a molecule can move, and each motion contributes to the energy of the molecule. *Molecular motion is quantized* — that is, each motion can occur with only certain energies. A molecule may gain or lose energy in one of these motions by absorbing or emitting a photon of light. The photon has energy that matches the allowed energy gain or loss associated with the motion.

By measuring the energies of the photons absorbed by the molecule, we obtain its spectrum. The spectrum provides information about the motions of the molecule. We begin with an expression for the energy of a molecule, written as a sum of the energies of the various motions possible to it. Each motion will then be described, along with the region of the electromagnetic spectrum with which it interacts.

The total energy of a molecule may be written as follows:

$$E_{total} = E_{spin} + E_{translation} + E_{rotation} + E_{vibration} + E_{electrons} + E_{nucleus}$$

NUCLEAR MAGNETIC RESONANCE—ESPIN.

The nuclei of some (but not all) atoms behave as if they were rotating, or spinning, about an axis passed through them, much as a top spins about its central axis.

This is quite easily visualized in terms of the particle nature of the nucleus; it is less easily visualized in terms of the wave picture, which by now we know is another face of nature that is manifested in the atomic/molecular realm. Spin of nuclei is similar to electron spin.

Examples of atoms whose nuclei spin are hydrogen, fluorine, the 13 isotope of carbon, and the 15 isotope of nitrogen. The spin of the hydrogen nucleus, because hydrogen atoms occur very frequently in molecules and are therefore very useful in determination of structure. When placed in a strong magnetic field, the spinning motion of the hydrogen atom gives rise to two energy levels.

The spin can be promoted from the lower to the higher level by interacting with and absorbing energy from electromagnetic radiation in the radio frequency (RF) region of the spectrum. A typical radio photon has a frequency of 10^8 s^{-1}, a wavelength of 3 meters (about ten feet), and an energy of 6.6×10^{-26} Joules.

The separation between the two spin energy levels is a sensitive function of the molecular environment of the hydrogen atom in a molecule. Thus hydrogen atoms in different environments absorb photons of different energies.

By plotting the amount of energy absorbed by the spinning nuclei versus the frequency of the RF radiation applied to the molecule, we obtain the *nuclear magnetic resonance (NMR) spectrum* of the molecule.

This spectrum provides information about the chemical environment of the spinning hydrogen nucleus, and can be used to deduce the atomic bonding patterns in the molecule.

NMR has now been developed to such a level of sophistication that it can be used to determine the 3-dimensional structures of huge protein molecules in solution. Recently, nuclear magnetic resonance spectroscopy has found medical applications, in which context it is known as magnetic resonance imaging (MRI). MRI utilizes the spinning motions of atomic nuclei to provide a map of the internal structure of human body tissue.

We will now discuss the nature of NMR spectra in a systematic way, beginning with spectra of simple molecules, and gradually escalating in complexity.

As we proceed through the discussion, we will introduce some of the jargon used by practicing chemists when discussing NMR. As a first bit of jargon, we will henceforth refer to hydrogen atoms as *protons*. This is appropriate, because it is the nucleus of the hydrogen atom that is spinning, and this nucleus indeed consists of a single proton.

We begin by considering the NMR spectrum resulting from the molecule. The molecule has a single proton on the right (on carbon 1), and a pair of protons on the left (on carbon 3). Carbon atoms 1 and 3 are separated by a carbon atom attached to two chlorine atoms. The single proton on the right is attached to a carbon atom that is attached to 2 chlorine atoms. In contrast, each proton on the left is attached to a carbon atom that is attached to one chlorine atom and a second hydrogen atom.

The environments of the protons on carbon 1 and carbon 3 are clearly different. The energy level spacing of the two types of protons should thus be different, and two signals, one for each type of proton, should appear in the NMR spectrum.

Several aspects of this spectrum are important. First, the single sharp signal at exactly 0 ppm is not due to the subject molecule, but is instead due to a reference compound named *tetramethylsilane*, or TMS. By general agreement, the positions of NMR signals are reported relative to the position of the signal for TMS.

Second, the unit used to measure resonance position, ppm (parts per million), is not a unit of energy. It has no units at all. It may be regarded as a measure of the shift in position between the TMS resonance and a particular resonance in the compound of interest.

It is usually called the *chemical shift*. The chemical shift indicates the chemical environment in which the spinning nucleus is found, as we will see shortly. Ignoring the TMS signal, then, which is present in all NMR spectra, we see that there are indeed two signals. Interestingly, the sizes, or intensities, of the signals are different.

Spin-spin Splitting—A Closer Look

To this point, we have discussed the most basic aspects of nuclear magnetic resonance spectroscopy. We have acknowledged the existence of spin-spin splitting, but have not said much about its molecular origins, and have not explained the origin of the number of subsignals in a multiplet.

In this section we will pursue these matters in a bit more depth. As a basis for discussion, consider a molecule in which there are three protons on carbon 1 and two protons on the neighboring carbon atom 2.

Our rules tell us that the signal due to carbon-1 protons will be split to a triplet, and that the signal due to carbon-2 protons will be split to a quartet. Why a triplet? Why a quartet?

Why do the subsignals of the triplet have relative intensities 1:2:1, while those of the quartet have relative intensities 1:3:3:1? What is the significance of the *magnitude* of the splitting between the components of a multiplet?

Focus first on the splitting of the signal due to the carbon-1 protons. The spinning protons on carbon 2 behave like tiny magnets. The magnetic field is generated by the spinning negative charge, which is much like an electric current. The direction of the magnetic field generated by a carbon-2 proton depends on its direction of spin (clockwise or counterclockwise). Either direction is equally probable for each of the three carbon-2 protons. Given this, it is easy to appreciate that there are 3 possible combinations of spin directions for the 2 protons on carbon 2. Both can be spinning clockwise, in which case their magnetic fields add together and reinforce the large laboratory magnetic field.

One can be spinning clockwise and the other counterclockwise, in which case their magnetic fields cancel. This arrangement can occur in two different ways because either proton 1 or proton 2 can have clockwise spin. The arrangement of one spin "up" (cw) and the other "down" (ccw) is thus twice as probable as the arrangement with both spins "up". Finally, both protons can have spin "down", in which case their magnetic fields add together but oppose the applied laboratory field. The spin arrangements of the carbon-2 protons are perhaps better appreciated when presented pictorially. We conclude that the carbon-1 protons will feel three slightly different magnetic fields, one slightly larger than, one equal to, and one slightly smaller than the applied field. The middle possibility is twice as probable as the other two.

Now we are in a position to appreciate why the signal due to the carbon-1 protons is split to 3 subsignals of relative intensities 1:2:1. This is a simple consequence of the fact that the carbon-2 protons create 3 slightly different magnetic fields that can be experienced by the carbon-1 protons. One signal arises from each of these magnetic fields, with intensity proportional to the probability of the spin arrangement giving rise to the magnetic field perturbation caused by protons on carbon 2. In similar fashion, we can understand why the signal due to carbon-2 protons is split to 4 subsignals.

The 3 protons on carbon 1 can have their spins aligned in 4 different ways (all "up" (one possibility), two "up" one "down" (3 possibilities), one "up" two "down" (3 possibilities), and all "down" (again only one possibility).

The carbon-2 protons will experience 4 slightly different magnetic fields with probabilities 1:3:3:1, so will be split to a quartet whose components have intensities reflecting these probabilities. The number of components in a multiplet and their relative intensities are thus seen to have a physically plausible explanation.

In general, as we have seen, "n" neighboring protons cause a splitting into "n+1" components. For reasons that we need not pursue, the relative intensities of the components resulting from splitting by any number of equivalent neighboring protons can be readily computed from Pascal's Triangle, in which a particular number is calculated as the sum of the numbers to the left and right of it in the row above.

We conclude our examination of spin-spin splitting by considering the magnitude of the separation between two adjacent signals within a multiplet. We can base the initial part of the discussion on the same molecule used above, in which there are three protons (call them protons A) on carbon 1 and two protons (protons B) on the neighboring carbon atom 2.

The magnetic interaction between protons A and B is called spin-spin coupling. Protons A are said to couple with (meaning that they interact with) protons B. The strength of the interaction determines the magnitude of the splitting between adjacent components of a multiplet.

The stronger the interaction, the larger the splitting. Further, spin-spin coupling between the two sets of protons A and B obeys a version of Newton's third law, that for every action there is an equal and opposite reaction.

If protons A interact with protons B with a certain strength, splitting the B signal by a certain amount, then protons B interact with protons A with exactly the same strength, splitting the A signal by exactly the same amount.

INFRARED SPECTROSCOPY—$E_{VIBRATION}$

Infrared spectroscopy allows us to examine the vibrational motions of molecules. We will discuss these motions as one of several types of motion that the molecule as a whole can undergo.

$E_{translation}$, $E_{rotation}$, $E_{vibration}$

The nuclear spinning motion discussed above involves only one atom of a molecule. The three motions to be discussed now — translation, rotation, and vibration — are motions of the molecule as a whole in space, involving concerted movements of all of the atoms.

For illustration, we use a very simple molecule consisting of only 2 atoms. Such molecules are called *diatomic molecules.* Examples are,

$$O_2, H_2, HCl, \text{ and } CO.$$

First, the entire molecule may move through space in an arbitrary direction and with a particular velocity. This is called *translational motion* and with it we associate the translational kinetic energy of the molecule:

$E_{trans} = 1/2\ mv^2$ (m = mass of molecule; v = velocity of centre of mass of the molecule)

The translational velocity of the molecule has components along each axis of a Cartesian coordinate system so,

$$E_{trans} = 1/2\ mv^2 = 1/2\ mv_x^2 + 1/2\ mv_y^2 + 1/2\ mv_z^2$$

where v_x is the x-component of velocity, etc., and m is the mass of the molecule

The total translational KE of the molecule is made up of three parts, each representing the kinetic energy of the molecule along one reference direction. The molecule has 3 translational degrees of freedom (i.e., independent directions in which it can translate), one corresponding to each Cartesian axis. A molecule travelling through space at a temperature of 300 K has translational kinetic energy of 6.2×10^{-21} Joules. However, the spacing between the allowed translational energies for an HCl molecule in a container 20 cm in length is on the order of only 10^{-40} J. This is such a small energy that transitions between translational kinetic energy levels are difficult to observe. At the present time, no spectroscopy is done involving these energy levels.

Second, the molecule may rotate about an internal axis. For example, rotation of a diatomic molecule is diagrammed in Figure. Like translation, a molecular rotation may be resolved into three mutually perpendicular components. The rotational kinetic energy of the molecule is,

$$KE_{rot} = 1/2\, I_x w_x^2 + 1/2\, I_y w_y^2 + 1/2\, I_z w_z^2,$$

where I_x, I_y, and I_z are the moments of inertia (the moment of inertia is the rotational equivalent of mass) about the x, y, and z axes, and w_x, w_y, w_z are the angular velocities (in units of radians/s) about these axes.

There are 3 *rotational degrees of freedom,* one for each Cartesian axis. An exception arises in the case of a linear molecule, for which one of the three axes is coincident with the molecular axis. The molecule has no rotational energy about this axis, because the corresponding moment of inertia is zero (i.e., no mass lies off the axis). Linear molecules have only two rotational degrees of freedom.

Rotation is a type of oscillation, in which the number of rotations per second is the frequency. If the molecule has associated with it a permanent electric field, then rotation of the molecule about some axis causes an oscillation in the electric field.

If this oscillation matches in frequency the oscillation in the electric field of an impinging photon, the molecule can absorb the photon and go to a higher energy rotational state. Rotations of molecules at room temperature typically occur with energy on the order of 10^{-23} J.

Photons with frequencies near 10^{11} s^{-1} (wavelength, 0.03 m), in the microwave region of the electromagnetic spectrum, are required to interact with such motions. The rotational (or microwave) spectrum of a molecule gives information about the moments of inertia, from which can be extracted the bond lengths and bond angles in the molecule. Interpretation of rotational spectra is complex, and we will not be concerned with it at this time.

Functional Group Analysis and Fingerprinting

An *organic functional group* is a small group of atoms, bonded together in a characteristic way, that occurs in a large variety of compounds. Examples are C=O (carbonyl), C(=O)OH (carboxyl), and NH_2 (amine).

The functional group is often emphasized in writing the formula of a compound containing it. Thus R-COOH represents any molecule containing the carboxyl functional group. R is a generic symbol for an organic "radical", consisting of primarily carbon and hydrogen atoms bonded together, and in which one carbon atom is capable of forming one additional bond.

Table. Functional Groups

Functional Group	*Wavenumber Range*
O-H	3400-3200
N-H	3300-3000
C(3)C-H	3300
=C-H	3100-3000
C(=O)-H	2700
C=O	1700
C=C	1680-1600
Benzene	1600
C-C-O	1260-1000
strong bands characteristic of benzene	900-650

Organic functional groups differ from one another both in the strength of the bond(s) and in the masses of the atoms involved. For instance, the O-H and C=O functional groups contain atoms of different masses connected by bonds of different strengths. According to equation, we therefore expect the O-H and C=O groups to absorb IR radiation at different positions in the spectrum. The presence of a strong, broad band between 3200 and 3400 cm^{-1} indicates the presence of an O-H group in the molecule, while the presence of a strong band around 1700 cm^{-1} confirms the presence of a C=O group.

For organic molecules, the infrared spectrum can be divided into three regions. Absorptions between 4000 and 1300 cm^{-1} are primarily due to specific functional groups and bond types. Those between 1300 and 900 cm^{-1}, the *fingerprint region*, arise from more complex vibrations in the molecules; and those between 900 and 650 cm^{-1} indicate benzene rings in the molecule. We will briefly discuss the characteristic infrared absorption frequencies of a variety of common organic functional groups. While studying these, realise that the infrared spectrum of even a simple compound can be very complex. Some absorption bands are weak while others are strong. Some are close enough together to coincide or overlap. Interactions such as hydrogen bonding cause absorption bands to shift and broaden.

Accordingly, do not expect to be able to identify every absorption band in a spectrum. Rather, look for the presence or absence of absorption bands in their characteristic spectral regions to confirm the presence or absence of particular functional groups.

Alkanes. Alkanes are hydrocarbons containing only single bonds. The most prominent bands in infrared spectra of alkanes and from alkane portions of more complicated organic compounds are due to C-H stretching and bending. The symmetric and asymmetric C-H stretching frequencies are in the regions 3000 to 2900 and 2900 to 2800 cm^{-1} respectively, and are usually weak. Because most organic compounds contain several $-CH_3$, CH_2, and/or -CH groupings, there is often overlap of absorption bands in this region. Thus the presence or absence of bands is taken to indicate the presence or absence of C-H bonds in the molecule.

The major bending modes of CH_2 and CH_3 groups appear at 1470-1420 and 1380-1340 cm^{-1}. The exact positions depend upon the nature of adjacent atoms. Interpretation is usually complicated by the presence of several bands of this type, or additional bands from other sources. The (usually) large number of C-C bonds in an organic molecule makes the C-C stretching vibrations in the 1300-1100 cm^{-1} region uninterpretable in most cases.

Alkenes. Alkenes are hydrocarbons containing at least one C=C double bond. When H is attached to a double-bonded C, C-H stretching bands generally appear in the region 3100-3000

cm^{-1}. Alkanes and alkenes can thus be differentiated. The C=C stretching frequency occurs in the 1675-1600 cm^{-1} region and varies with the atoms attached to the double-bonded carbons.

Table. Alkene Bending Absorptions

Alkene	*Wavenumber range, cm^{-1}*
R-CH=CH_2	1000-960 940-900
$R_2C=CH_2$	915-870
trans-RCH=CHR	990-940
cis-RCH=CHR	790-650
R_2C=CHR	850-790

Alkynes (or *acetylenes*) Alkynes are hydrocarbons containing a carbon-carbon triple bond, C(3)C). The C-H stretching vibration of *terminal acetylenes* (meaning that the last carbon in a chain is triple bonded to its neighboring carbon) generally appears at 3300 cm^{-1} as a strong sharp band.

The CC stretching band is found at 2150-2100 cm^{-1} if the alkyne is monosubstituted (i.e., R-C(3)C-H) and at 2270-2150 cm^{-1} if disubstituted (i.e., R-C(3)C-R′). These absorptions are usually weak. Aromatic (benzene) Rings. Aromatic C-H stretching absorption appears in the region 3100-3000 cm^{-1} and is readily distinguished from alkane, alkene, or alkyne C-H, allowing a reliable determination of the types of carbon-bound hydrogen in the molecule. The positions of aromatic C-H bending bands at 900-690 cm^{-1} depend upon the substitution pattern of the benzene ring as indicated in Table.

Table. Absorptions of Benzene Rings

Substitution	*Wavenumber range (cm^{-1})*
mono	755-730 710-690
1,2-di	765-730 710-690
1,3-di	800-750 710-690
1,4-di	840-800
1,2,3-tri	800-760 740-700
1,2,4-tri	880-860 820-800

In the absence of other interfering absorptions, these strong, usually sharp bands can be used to distinguish postitional isomers of substituted benzenes. Sharp bands at ~1600 and ~1500 cm^{-1} are characteristic of all benzene compounds; a band at 1580 cm^{-1} appears when the ring is *conjugated* with a substituent. (The word conjugated means the substituent has a double bond one bond removed from the benzene ring).

Alcohols. Alcohols are compounds containing the O-H functional group. The characteristic infrared band due to the O-H stretching vibration appears at 3650-3600 cm^{-1} in dilute solution. With increasing concentration in solution, or in pure liquids or solids, intermolecular hydrogen bonding broadens the band and shifts its position to lower wavenumber (3500-3200 cm^{-1}).

The infrared spectrum of pure t-butyl alcohol shows the broad O-H stretching band at 3440 cm^{-1}. In addition to the O-H stretching vibration, alcohols exhibit O-H bending and C-O stretching transitions at 1500-1300 cm^{-1} and 1220-1000 cm^{-1}, respectively. Although C-O stretching bands occur in a spectral region where there are usually many other bands, they are easy to identify because they are intense (strong). They are often *coupled to* (i.e., move in concert with) C-C absorption bands and therefore exhibit splitting.

The exact location of the C-O bands depends on the degree of branching of the carbon atom that is attached to oxygen. Thus, as Table reveals, the position of the C-O absorption band can allow distinction between *primary, secondary,* and *tertiary* alcohols.

In a primary alcohol, the OH group is attached to a carbon that is in turn attached to only one other carbon atom. In secondary and tertiary alcohols, the OH group is located on a carbon attached to 2 and 3 other carbons, respectively.

Ethers. Ethers are compounds containing the C-O-C functional group. The asymmetric C-O stretching absorption of ethers appears in the region 1280-1050 cm^{-1}. As in alcohols, the position of this strong band is dependent on the nature of the attached groups.

Table. C-O Stretching Vibration of Alcohols

Primary alcohols	1050 cm^{-1}
Secondary alcohols	1125 cm^{-1}
Tertiary alcohols	1200 cm^{-1}

Aldehydes and Ketones. The carbonyl, (C=O) stretching absorption is particularly important because it is present in a variety of functional groups including aldehydes, ketones, and carboxylic acids and their derivatives. Table shows the basic structures of compounds containing C=O.

Table. Carbonyl Containing Compounds

Name	*Structure*
Aldehyde	R-C(= O) – H
Ketone	R-C(=O) – R
Carboxylic Acid	R-C(=O) – OH
Anhydride	R-C(=O)–O –C(=O) – R
Ester	R-C(=O)–OR

The C = O stretching frequencies of aldehydes in which R is an alkane are observed at 1735-1710 cm^{-1}. Although the position of the carbonyl absorption does not allow distinction between aldehydes and ketones, the former are recognizable by the C-H stretching vibration that appears as two bands in the 2850-2700 cm^{-1} region.

Double bonds in conjugation with the carbonyl group lower the wavenumber by 25 to 50 cm^{-1}. Thus, aromatic aldehydes (benzene-C(=O)H) generally absorb at 1700-1690 cm^{-1} and diaromatic ketones at 1670-1660 cm^{-1}. Intramolecular hydrogen bonding to the carbonyl oxygen also lowers the wavenumber by 25 to 50 cm^{-1}.

Carboxylic Acids. Carboxylic acids contain the C(=O)OH functional group. The carbonyl group of acids in which R is not aromatic appears at 1730-1700 cm^{-1} and is shifted to 1720-1680 cm^{-1} when the C=O double bond is conjugated with a C=C double bond. The most characteristic absorption of carboxylic acids is a broad band from 3300-2500 cm^{-1} due to hydrogen bonded O-H stretching. The C-H stretching vibrations appear as small bands on top of this band.

Amines. Amines are compounds containing the N atom singly bonded to R or H. Primary amines have formula NRH_2, while secondary and tertiary amines have formula NR_2H and NR_3 respectively.

Primary and secondary amines show N-H stretching vibrations in the 3500-3300 cm^{-1} region. Primary amines generally have two bands approximately 70 cm^{-1} apart due to asymmetric and symmetric stretching modes. Secondary amines show only one band.

Inter- or intramolecular hydrogen bonding broadens the absorptions and lowers the frequency. In general the intensities of N-H bands are less than of O-H bands. The N-H bending and C-N stretching absorptions are not as strong as the corresponding alcohol bands and occur at approximately 100 cm^{-1} higher frequencies. NH_2 groups give an additional broad band at 900-700 cm^{-1} due to bending.

Nitriles. Nitriles are compounds containing the C(3)N group. A sharp, unusually strong absorption at 2260-2220 cm^{-1} due to CN stretching is characteristic of nitriles. To complete the section on IR spectroscopy, we make one broad and extremely useful generalization.

Every compound, organic or inorganic, has a unique infrared spectrum. This spectrum serves as a fingerprint for the compound. Thus if a compound prepared in the laboratory is thought to be a particular known substance, the fastest way to determine whether or not it is is to obtain the IR spectra of the compound and the known substance. If the spectra are identical (band-for-band match), then so are the substances.

ELECTRON MOTION IN MOLECULES

The electrons can change energy states within the molecule. When they do this, they change location with respect to the atomic nuclei. Motions of electrons within the molecule are associated with frequencies that are typically in the range 10^{14} to 10^{15} s^{-1}.

When this frequency matches the frequency of light impinging on the molecule, the light can be absorbed, and an electron moves to a higher energy state. The frequencies of light absorbed give us information about the motions of the electrons in the molecule. Photons with frequency 10^{15} s^{-1} have energy of 6.6×10^{-19} J, and fall in the visible and ultraviolet regions of the electromagnetic spectrum. Spectroscopy involving changes in electron motions is therefore commonly called UV-Visible spectroscopy. It is also called electronic absorption *spectroscopy* (EAS).

The wavelengths of light absorbed and the extent of absorption by a substance can be measured using an instrument called an electronic absorption spectrometer. A typical spectrometer covers the wavelength range between 1100 and 190 nm and thus ranges from the near IR region (1100-750 nm) to the near UV region (400-190 nm) ("near" means bordering on the visible region). To measure the UV-Vis spectrum of a substance requires the preparation of a solution of the substance in an appropriate solvent.

The spectrum is a plot produced by the spectrometer of the amount of light absorbed (the absorbance) as a function of the wavelength of the light. The spectrum of potassium perm anganate, $KMnO_4$, dissolved in water. Note that the absorption does not consist of sharp lines, as in NMR, or even as relatively narrow signals, as in IR. Instead, absorption usually occurs over a range of wavelengths. Each absorption region is therefore called an *absorption band,* to indicate that it covers a band, or range, of wavelengths, rather than occurring at just a single wavelength.

The amount of light absorbed by the substance — the absorbance — at a particular wavelength is directly proportional to 1) the length of solution through which the light beam passes; and 2) the concentration of substance (number of molecules or moles per unit volume) in the solution.

$$A = e\,l\,M$$

where A = absorbance (amount of light absorbed); l = the length of solution through which the light passes, in cm; M = the molarity (number of moles of substance per liter) of the solution; and e = the proportionality constant relating A to l and M. e is called the molar absorptivity of the substance at the particular wavelength at which A is measured. The most important aspect of this equation for us at this point is this: the amount of light absorbed is directly proportional to the amount of substance per unit volume of solution. This is called Beer's Law.

The molar absorptivity is characteristic of the substance and the wavelength. Values of e for a substance are usually quoted for the wavelengths of maximum absorbance, l_{max} — that is, the tops of the absorption bands. The value of e gives a measure of the effectiveness of a molecule of substance at capturing a photon of light. Values can range from as low as 0.01 to as high as 10^5 L/mole-cm. Note that substitution of 1000 cm^3 for 1 L reduces the units of e to cm^2/mole.

This shows that e measures the effective cross-sectional area of the substance with respect to capturing light. Since the e values of a substance are characteristic of it, we may use the amount of light absorbed to measure the concentration of the substance in solution.

Example: The spectrum of a mixture of $KMnO_4$ and $K_2Cr_2O_7$ dissolved in water has A = 0.266 at 565 nm and A = 0.790 at 350 nm. For $KMnO_4$, e = 1.13×10^3 at l = 350 nm and 1.27×10^3 at 565 nm. For $K_2Cr_2O_7$, e = 2.84×10^3 at l = 350 nm and essentially zero at 565 nm. Calculate the concentrations of $KMnO_4$ and $K_2Cr_2O_7$ in the solution.

Solution: Since dichromate does not absorb at 565 nm, the absorbance at 565 is due entirely to permanganate. It's concentration can be calculated from Beer's Law:

$$[MnO_4^-] = \text{A at 565/e at 565} = 0.266/1.27 \times 10^3$$
$$= 2.09 \times 10^{-4}\ \text{M}$$

The absorbance at 350 nm is a sum of contributions from the two species; however, the permanganate contribution can be calculated and subtracted, allowing the dichromate concentration to be obtained:

$$A_{MnO4-} \text{ at } 350 = (\text{e at } 350) \times [MnO_4^-]$$
$$= 1.13 \times 10^3 \times 2.09 \times 10^{-4}$$
$$= 0.236$$
$$A_{Cr2O72-} \text{ at } 350 = A_{total} - A_{MnO4-}$$
$$= 0.790 - 0.236 = 0.554$$

$$[Cr_2O_7^{2-}] = (A \text{ at } 350)/(e \text{ at } 350)$$
$$= 0.554/2.84 \times 10^3$$
$$= 1.95 \times 10^{-4} \text{ M}$$

Example: The complex ion, $FeL(CH_3CN)(CO)^{2+}$ has the structure shown in Figure. L represents the cyclic structure consisting of carbon, hydrogen, and nitrogen atoms. The complex decomposes when dissolved in acetonitrile, CH_3CN, according to the following equation

$$FeL(CH_3CN)(CO)^{2+} + CH_3CN \text{ ® } FeL(CH_3CN)_2^{2+} + CO(g)$$

The course of this reaction can be monitored by measuring the absorbance of the solution at 556 nm, a wavelength where $FeL(CH_3CN)_2^{2+}$ absorbs but $FeL(CH_3CN)(CO)^{2+}$ does not. The following absorbance data are obtained at various times after the reaction begins:

Time, minutes	*Absorbance*
0	0
3	.121
8.5	.243
11.5	.302
14.5	.369
17	.397
20	.447
24	.487
28	.528
36.5	.579
45	.615
59.5	.646

The initial concentration of $FeL(CH_3CN)(CO)$ was 0.694×10^{-4} M. The value of e for $FeL(CH_3CN)_2^{2+}$ at 556 nm is 9650 $M^{-1}cm^{-1}$.

- Plot $[FeL(CH_3CN)(CO)^{2+}]$ vs time.
- What percentage of the reactant remains after 45 minutes?
- What is the average rate of the reaction over the first 15 minutes?

Solution: The concentration of $FeL(CH_3CN)(CO)^{2+}$ at each time must be calculated from the measured absorbance, the molar absorptivity of the product, $FeL(CH_3CN)_2^{2+}$, and the known initial concentration of $FeL(CH_3CN)(CO)^{2+}$. For example, at t = 3 minutes, the measured absorbance is 0.121.

The corresponding $[FeL(CH_3CN)_2^{2+}]$ *is:*

$[FeL(CH_3CN)_2^{2+}]$ = A/e = 0.121/9650 = 1.25×10^{-5} M.

The reaction stoichiometry indicates that 1 mole of product is obtained from 1 mole of reactant. It follows that $[FeL(CH_3CN)(CO)^{2+}]$ = (6.94–1.25) × 10^{-5} = 5.69×10^{-5} M. An analogous calculation can be performed at each time in the table. This enables a plot of $[FeL(CH_3CN)(CO)^{2+}]$ versus time to be constructed. Find the concentration of $FeL(CH_3CN)(CO)^{2+}$ from the curve at t = 45 minutes. Divide this by the initial concentration of reactant to find the fraction remaining at 45 minutes. Then multiply by 100.

When t = 45 min, the curve indicates that $[FeL(CH_3CN)(CO)^{2+}]$ has fallen to 5.7×10^{-6} M. The fraction of reactant remaining is thus:

$$5.7 \times 10^{-6}/6.94 \times 10^{-5} = 0.082, \text{ or } 8.2 \text{ per cent}$$

Divide the change in $[FeL(CH_3CN)(CO)^{2+}]$ that occurs over the first 15 minutes by the elapsed time, 15 minutes. This gives the average change in concentration per minute over this time interval:

$$(6.94 - 3.05) \times 10^{-5} \text{ M/15 min} = 2.59 \times 10^{-6} \text{ M/min}$$

Example: The electronic absorption spectrum of a water solution of $CrCl_3$ (chromium(III) chloride) is shown in Figure. The absorption bands, resulting from the Cr^{3+} ion, are in the visible region of the spectrum, and are responsible for the violet colour of the solution. What are the energies of the most strongly absorbed photons? Plot the electronic energy levels of the Cr^{3+} ion on an energy level diagram.

Solution: The most strongly absorbed photons have energies corresponding to the wavelengths at the tops of the absorption bands. The energies, read directly from the absorption maxima in the spectrum, are at 16500 and 23000 cm^{-1}. These are in units of reciprocal wavelength and must be converted to energy units using the relationship,

E = hn = hc/lFor the 16500 cm^{-1} photon, E = (6.626 × 10^{-34} J/s)(3.0 × 10^{8} m/s)(16500 cm^{-1})(100 cm/m) = 3.28×10^{-19} JFor the 23000 cm^{-1} photon, E = 4.57×10^{-19} J, by the same method. To construct the energy level diagram, we must realise that the observed absorptions correspond to transitions of an electron in the Cr^{3+} ion from the lowest energy state to two higher energy states, one corresponding to each absorption band. The photon energies calculated above give us the gaps between the energy levels.

The energy level diagram should thus show three states. The middle state should be 3.28×10^{-19}J above the lowest one. The highest state should be 4.57×10^{-19} J above the lowest one. The diagram is shown in Figure.

Correspondence of Colour and Spectrum

Many substances are colored, resulting from absorption of light in the visible region of the spectrum. In addition to making these substances visually interesting, colour is an important physical property, and is used with such quantities as melting point and boiling point in describing the substance. It is important to have a qualitative understanding of the relationship between the colour of light absorbed by a substance and the colour of light that we perceive when we look at the substance.

These colors are *complementary*. We see the light that is NOT absorbed by the colored substance. The colour that we see is that of the light that is not absorbed. Although this statement may contradict what you have previously thought, it nonetheless makes sense.

Photons that are absorbed by a colored substance cannot reach the eye to be observed. The photons reaching the eye are those that are NOT absorbed.

White light is made up of the three primary colors, red (R), green (G), and blue (B). The relationships between and combinations of these three colors are best shown in the form of the colour wheel. The wheel is set up with the primary colors at 12, 4, and 8 of a clock face. The complements of the primaries are shown directly opposite the corresponding primary. Thus cyan, the complement of red, is at 6 oclock. Cyan, C, is a combination of equal amounts of B and G, which occur to either side of it. Similarly, magenta (R + B) occurs opposite its complement, green; and yellow (R + G) appears opposite its compement, blue. A substance that absorbs light of a particular colour appears to have the colour opposite the absorbed colour on the wheel.

Thus, if red is absorbed, cyan is seen. Figure shows the correspondence between colour and wavelength of light, which may be seen by viewing white light through a simple device called a *spectroscope.*

The colour of a substance may be understood in terms of its electronic absorption spectrum. Thus if a substance has an absorption band at 550 nm, in the green region of the spectrum, it absorbs green photons, transmits red and blue, and should appear to us to be magenta (R + B).

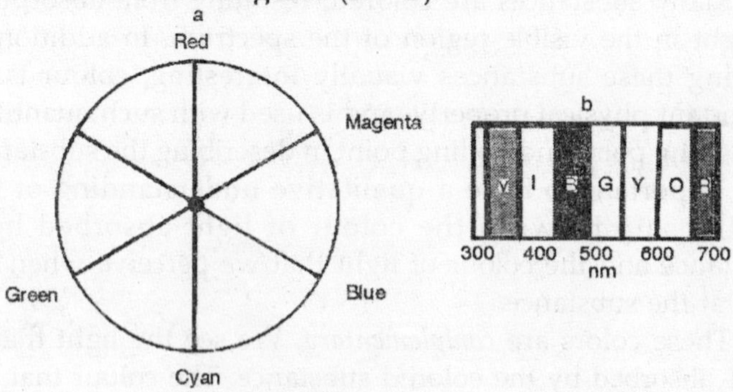

Fig. Colour

Relationship of Structure

It is important to have some idea of the types of molecular electronic structural features that are likely to give rise to absorption bands in the near-IR (1100-750 nm), visible (750-400 nm), near-UV (400-190 nm), and far UV (shorter wavelength than 190 nm) regions of the electromagnetic spectrum. We will make some broad statements for which you can expect many exceptions.

The purpose of the statements is to enable you to make reasonable predictions about electronic transitions in molecules. In making predictions, of course, we run the risk of being wrong. But an incorrect prediction based on sound reasoning is far preferable to no prediction (and no reasoning) at all. You are thus encouraged to be fearless in making predictions.

The near-IR and Visible regions (1100-400 nm). Several structural features are notable for generating light absorption in these regions. First, compounds containing *transition metal ions* (ions of the metals found in the d block of the periodic table) show an interesting rainbow of colors.

These ions are known for their ability to form *complexes,* compounds in which up to six molecules or ions are covalently bonded to the transition metal ion. In this context, the molecules or ions are called *ligands*. In theory, any molecule or ion with at least one lone (nonbonding) electron pair can serve as a ligand by using its electron pair to bond to a transition metal ion. In practice, molecules containing oxygen (2 lone pairs), nitrogen (1 lone pair), sulfur (2 lone pairs), phosphorus (1 lone pair), or halogen (3 lone pairs) frequently serve as ligands. Table gives formulas and colors for a collection of complexes containing the Co^{3+} ion.

Table. Complexes of Co^{3+}

Complex	*Color*
$[Co(NH_3)_6]Cl_3$	yellow
$[Co(NH_3)_5Cl]Cl_2$	purple
$[Co(NH_3)_4Cl_2]Cl$	green
$[Co(NH_3)_4Cl_2]Cl$	violet

Without going into too much detail, the colors of transition metal ions are a consequence of the partial filling of the d valence orbitals with electrons. When ligands bind to the transition metal, the d orbitals are caused to have different energies, so that the d electrons can undergo light-promoted transitions between them.

The energy spacing between the d orbitals is influenced by the chemical identity of the ligands; thus different ligands coax different colors from the same metal ion. These so-called *d-d transitions* are typically quite weak, with molar absorptivities in the range 0.1-100 $M^{-1}cm^{-1}$ Second, organic molecules with extended *conjugation* (alternating single and double bonds) often absorb visible light.

Acid-base indicators have structures of this type. Extended conjugation normally leads to quite intense colour (strong absorbance, large molar absorptivity), in contrast to transition metal complexes, whose colors are relatively pale.

The *porphyrin ring*, which occurs in many biologically-important molecules, has benzene-like resonance involving its double bonds. It absorbs intensely in the visible region. Electronic absorption spectral data for a number of compounds of this type are presented in Table.

Table. Spectral Data for Conjugated Species

Molecule	*Absorption Wavelengths and Molar Absorptivities,* $\lambda_{max}(\varepsilon)$
2-nitrophenol, $C_6H_4(OH)(NO_2)$	279 (6600) 351 (3200)
2-nitroaniline, $C_6H_4(NH_2)(NO_2$	283 (5400) 412 (4500)
Acetophenone, $(C_6H_5)(CH_3C{=}O$	240 (13000) 278 (1110 319 (50)

Third, species (molecules or ions) in which an atom in a high oxidation state is bonded to species with lone pairs of electrons (such as oxide ions and halide ions) are often colored, due to the transition of an electron from one of the lone pairs to the electron-deficient high-oxidation-state atom.

Such transitions are called *charge transfers*, because they essentially transfer negative charge (an electron) from a region of negative to a region of positive charge.

Table. Spectral Data for Charge Transfer Absorptions

Species	*Absorption Wavelengths*
MnO_4^-	350 (1130) 565 (1270)
$Fe(CN)_5(C_5H_7N_2)^{2-}$	660 (10^4)

The permanganate ion, MnO_4^-, is a good example of a species whose colour is due to a charge transfer transition. The anion is intensely purple both in the solid state and in aqueous solution. Table gives spectral data for charge transfer absorptions in a number of species.

Near UV region (400-190 nm). Many organic molecules with double bonds, particularly those in which one of the double bonded atoms has at least one lone pair, absorb radiation in the near UV region.

Examples include acetone, with a carbon-oxygen double bond and 2 lone pairs on oxygen; benzene, with 3 conjugated double bonds; acetonitrile, with a carbon-nitrogen triple bond and a lone pair on nitrogen; and SO_4^{2-} (an inorganic ion), with sulfur-oxygen double bonds and lone pairs on oxygen. Table presents absorption data for a number of common molecules of this type.

Table. Spectral Data for Species Absorbing in the Near-UV Region

Molecule	*Absorption Wavelengths and Molar Absorptivities, $\lambda_{max}(\varepsilon)$*
Acetone $(CH_3)_2C{=}O$	279 (13)
Nitromethane CH_3NO_2	275 (15)
Phenol, C_6H_5OH	210.5 (6200) 270 (1450)
Aniline	230 (8600) 280 (1430)

Far UV region (< 190 nm). Molecules containing strong sigma bonds and no lone pairs tend to absorb in this region. The only promotable electrons are those forming the single bonds, and they must receive a large amount of energy in order to undergo a transition to the lowest energy empty orbital of the molecule.

Table. Spectral Data for Species Absorbing in the Far UV Region

Molecule	*Absorption Wavelengths and Molar Absorptivities, $\lambda_{max}(\varepsilon)$*
Methanol, CH_3OH	177 (500)
Diethyl ether, $(C_2H_5)_2O$	188 (1995) 171 (3981)
Water	167 (7000)
Ethane, C_2H_6	135

Thus molecules such as ethane, C_2H_6 and other

hydrocarbons have no absorptions at wavelengths longer than 190 nm.

Strongly single bonded molecules with lone pairs on electronegative atoms also tend to absorb only at very short wavelength (high energy).

Thus water, HF, NH_3, and the like have no absorptions in the visible or near-UV regions of the spectrum. Table shows some typical far-UV data.

An Intuitive Development of Beer's Law

It is possible to arrive at Beer's Law using a simple macroscopic physical system consisting of an inclined plane perforated with holes at a certain density, and divided into four sections of equal length. The plane is pictured in Figure.

Suppose that 10000 identical small steel spheres are released at the top of the plane, and that we observe that 9000 arrive at the end of the top section, having avoided falling into one of the holes. Based on this information, how many spheres will arrive at the bottom of the plane? Two approaches seem reasonable on the surface. The first is to assume that the same number of spheres will fall into holes in each section. If 1000 fell in the first section, then a total of 4000 will fall along the entire length of the plane, leaving 6000 to reach the bottom.

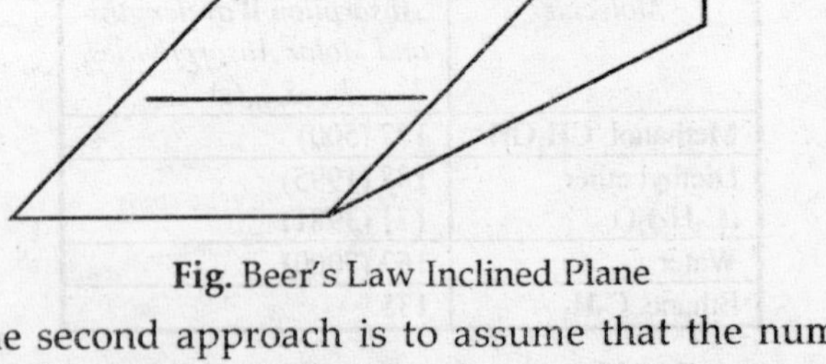

Fig. Beer's Law Inclined Plane

The second approach is to assume that the number of spheres that fall into holes in a given section is proportional to the number that enter the section; i.e., that the same fraction of spheres will disappear in each section. By this reasoning, we predict that 8100 spheres will survive the first 2 sections, 7290 will survive through sections, and 6561 will reach the bottom. This second line of reasoning proves correct: the number of balls that survive a section is proportional to the number of balls that start into the section. We write this in equation form:

$$N_n = k \times N_{n-1}$$

This equation states that the number of spheres arriving at mark n is proportional to the number of spheres that leave the preceding mark, n-1. k is a proportionality constant that must have a value between 0 and 1; it has the value 0.9 for our example above.

We can also express the equation in terms of the change in the number of balls:

$$DN = N_n - N_{n-1} = k \times N_{n-1} - N_{n-1} = N_{n-1} \times (k-1) = -K \times N_{n-1}$$

Note that because $k < 1$, the quantity k-1 is negative, consistent with the fact that balls are being lost and therefore that the change in the number of balls defined in equation is also negative. We have therefore replaced k-1 with –K, where K is a new, positive, constant.

Here are two questions that we now consider: How does the value of K depend on the length of the interval, l ? How does the value of K depend on the density of holes, r ? Intuitively, we expect that a longer interval will result in more balls being lost. Thus K increases as l increases. Also, we expect more balls to be lost if the density of holes is increased. So K increases as r increases.

In equation form, this becomes:

$$K = \zeta \times l \times r$$

where z is a constant representing the number of balls lost per unit density of spheres per unit length of section. Substituting 4S-3 into 4S-2 gives:

$$\Delta N = -\zeta \times l \times \rho \times N$$

Assuming an interval of infinitesimal length will result in an infinitesimal loss of spheres:

$$dN = -z \times r \times N \times dl$$

Separating the variable, N, from the variable, *l*, gives:

$$dN/N = -\zeta \cdot \rho \times dl$$

Finally, integration provides:

$$\ln(N/N_o) = -\zeta \times \rho \times l$$

Translating to base-10 logarithms gives:

$$\log(N/N_o) = -(1/2.303) \times \zeta \times \rho \times l = -e \times \rho \times l$$

Of course this equation can be expressed in exponential form as $N/N_o = 10^{-e \times r \times l}$. In words, this equation states that the fraction of balls surviving an interval decreases exponentially with the length of the interval and the density of holes in the interval. Let's now make a direct analogy between our inclined plane and a solution of a substance that absorbs light at some wavelength. We think of *l* as the length of solution through which the light beam must pass; r as the density (concentration) of absorbing molecules in the solution; the small spheres as photons of light; and N/N_o as the fraction of light transmitted by the solution:

$$\log(I/I_o) = -e \times M \times l$$

We have replaced the density of holes with the concentration of the solution in moles absorbing substance per liter, and the number of spheres with the intensity, I, of light. The intensity is simply the number of photons per second arriving at a particular location, so is a close analog of N. We now define a quantity called the absorbance, A, as $A = e \times M \times l$ so that equation becomes:

$$\log(I_o/I) = A.$$

A represents the amount of light absorbed (the number of spheres lost). Note that when all light is transmitted by a sample, $I_o/I = 1$ and $A = 0$. When 1/10 of the light is transmitted, $I_o/I = 10$ and $A = 1$. Thus A goes up as the fraction transmitted goes down. The equation $A = e \times M \times l$ is, of course, Beer's Law.

Chapter 7

The Gas Phase

INTRODUCTION

Gases and their properties play an ubiquitous and critical role in our daily lives. Life is supported by the oxygen of the air, which reacts with glucose in biochemical combustion. Living systems have evolved complex molecular architecture to absorb oxygen from the air; carry it to the cells; deliver it to the cells; transport it within the cell; and supervise its multistep, controlled reaction with glucose.

It is the rapid expansion of gases produced in the explosive combustion of gasoline that performs the work of the internal combustion engine. Expansion and contraction of gases in response to temperature and pressure changes is at the basis of continually changing weather patterns. Clearly an understanding of the properties, behaviour, and chemical reactivity of gases is an important foundation for developing an appreciation of complex biological and technological systems.

Following a discussion of the ideal gas law, a simple mathematical model for gas behaviour, we discuss the Kinetic Molecular Theory of gases, which establishes a link between molecular motion and temperature.

MACROSCOPIC PROPERTIES OF GASES

All gases have observable properties in common:

- Gases flow readily from one space to another and occupy all available space.
- Gases assume the shapes of their containers.

- Gases are readily compressible. By exerting force, we can cause a gas to occupy a smaller volume.
- Two or more gases form homogeneous mixtures (solutions) in all proportions. An example is air, a mixture of primarily nitrogen and oxygen with small amounts of other gases.
- Gases diffuse rapidly. This means that a gas can move across a large space in a relatively short time. Examples are the odors of perfume and skunk.

The state of a gas is described by the values of 4 macroscopic variables: the volume (V) occupied by the gas; the temperature (T) of the gas; the pressure (P) that the gas exerts on the walls of its container; and the amount of gas in moles (n).

The values of these quantities are related by a very simple equation called the ideal gas law. We will discuss this equation after a brief word on pressure.

Pressure

Pressure is defined as force per unit area. We illustrate the concept of pressure with an example.

Example: The bottom of each foot of a 150-pound man has a surface area of about 24 square inches. When the man stands on a bathroom scale, what pressure is exerted by his feet?

Solution: The force exerted by the feet is the man's weight, 150 lb. This is distributed over the area of the bottoms of both feet, or 48 square inches.

The pressure is then:

$$P = F/A = 150 \text{ lb}/ 48 \text{ in}^2 = 3.1 \text{ lb/in}^2$$

If the man stands on the scale with only one foot, the force is the same, but the pressure is doubled because the area is cut in half.

Gases exert pressure on the walls of their containers. Since a gas uniformly occupies a container, it exerts the same pressure on all the walls. The gases that make up the atmosphere exert pressure on the surface of the earth. This pressure, called the atmospheric pressure, can be measured using a barometer.

The left side of the figure shows a beaker or cup containing mercury. A glass tube of length 1 m and cross-sectional area, A, fitted with a stopcock on top, is inserted into the beaker so that the end of the tube is below the level of mercury in the beaker.

Initially, the stopcock is open and the levels of mercury inside and outside the glass tube are the same. A vacuum pump is then attached to the top of the glass tube, and the air is pumped out. As this occurs, the mercury level rises in the tube until it reaches a height, h. Despite further pumping, the mercury level rises no higher in the tube. When all air has been pumped out of the tube, the stopcock is closed, and the mercury maintains its position in the tube.

This behaviour is understood as follows. Initially, before pumping, the atmosphere exerts the same pressure on the surfaces of mercury inside and outside the tube. Since P_{out} = P_{in} = atmospheric pressure, the mercury levels inside and outside the tube are the same.

As air is pumped out of the tube, its pressure on the mercury inside the tube decreases, and mercury rises in the tube to a height, h, such that the force due to the mass of mercury in the tube just counterbalances the pressure of the atmosphere on the mercury outside the tube.

The height h is found to vary from day to day because the pressure of the atmosphere varies with weather conditions. However, its value is always in the neighborhood of 0.76 meters. For this reason, a unit of pressure called the atmosphere (atm) is defined as the pressure that will support a column of mercury 76 cm, or 760 mm, in height.

$$1 \text{ atm} = 760 \text{ mm Hg}$$

The pressure exerted by a column of mercury 1 mm in height is called a torr after Evangelista Torricelli, inventor of the barometer.

Thus:

$$1 \text{ atm} = 760 \text{ mm Hg} = 760 \text{ torr}$$

In the English and SI systems, the units of pressure are respectively the pound per square inch and the pascal (1 Pascal = 1 Newton per square meter, where a Newton is the force required to accelerate a 1 kg body at 1 m/s^2). The pressure in Pa exerted by a column of mercury 760 mm in height can be readily calculated and will provide a conversion factor between SI and traditional non-SI units.

The force exerted by a mass, m, of mercury is mg, where g is the acceleration due to gravity. This force is distributed over the cross sectional area, A, of the tube. The mercury therefore exerts the pressure, P = mg/A on the surface of mercury at the bottom of the tube. The mass of mercury can be calculated from its density, r, and its volume, which is Ah.

Substituting for m in the expression for pressure gives equation:

$$P = \rho \times (Ah)g/A = \rho \times gh$$

The pressure exerted by a column of mercury (or any liquid) of height h is independent of the cross sectional area of the column. The diameter of the barometer tube is therefore unimportant. Substituting the density of mercury (13.595 g/cm^3), the acceleration constant (g = 9.80665 m/s^2), and h = 0.760 m in 5–1–1 gives P = 1.0132×10^5 Nm^{-2} (1 Nm^{-2} = 1 Pascal).

The following conversion relations are thus obtained:

$$1 \text{ atm} = 14.7 \text{ lb/in}^2 = 1.013 \times 10^5 \text{ Pa } (1 \text{ Pa} = 1\text{N/m}^2)$$
$$= 1.013 \text{ bar } (1 \text{ bar} = 10^5 \text{ Pa})$$

Although the bar is official, the atmosphere and the torr are still the most commonly used pressure units in chemistry. We will therefore use the atm and the torr.

Note that atmospheric pressure is 14.7/3.1 = 4.7 times greater than the pressure exerted by the feet of the man in Example. The magnitude of the pressure exerted by the atmosphere is equivalent to having that 150-lb man, holding a 550-lb weight, standing on your stomach. The pressure of a gas confined in a container cannot be measured with a barometer. For this a variety of devices may be used, the simplest being the manometer. A manometer is a U-shaped glass tube containing mercury or some less toxic fluid.

The gas container is attached to one side of the manometer, and the other side is either open to the atmosphere (an open-end manometer) or evacuated (a closed-end manometer). The difference in the levels of fluid in the two arms of the manometer is a measure of the pressure of gas in the bulb through the relationship, $P_1 = P_2 + P$. Here P_1 is the pressure exerted by the gas on arm 1 of the manometer, P_2 is the pressure exerted on the arm of the manometer not connected to the gas container (arm 2), and P is the height of fluid in arm 2 minus the height in arm 1. P is positive if fluid is higher in arm 2 and negative if it is higher in arm 1.

For an open-end manometer, P_2 is atmospheric pressure; for a closed-end manometer, P_2 is zero. If the manometer fluid is mercury, the pressure is in torr and may be converted to atm using the conversion factor given above. If the manometer contains some other fluid, the height of fluid may be converted to an equivalent height of mercury using the densities of the fluid and mercury.

Example: An open-end manometer containing water as the working fluid is used to measure the pressure of gas in a bulb. The level of water in the side attached to the bulb is 8.65 cm higher than that in the side open to the atmosphere. Atmospheric pressure is 693 torr. What is the pressure of gas in the bulb?

Solution: Use the relation $P = r \times g \times h$. To find the height of mercury equivalent to 8.65 cm of water, we equate $r \times gh$ for water to $r \times gh$ for mercury.

The gravitational acceleration cancels, giving:

$$(1.00\ g/cm^3)(86.5\ mm) = (13.60\ g/cm^3)(h)h(Hg) = 6.4\ mm$$

The pressure of gas is this amount less than atmospheric pressure:

$$P_{gas} = 693–6.4 = 687\ torr.$$

THE IDEAL GAS LAW

The ideal gas law is a relationship between the pressure (P), the volume (V), the amount (n), and the temperature (T) of a gas. The ideal gas law is called an equation of state, because it represents all possible states — combinations of P,V and T — in which the amount of gas, n, can be found.

$$PV = nRT$$

In this equation, P is the pressure in atm; V is the volume in L; n is the amount of gas in moles, and T is the Kelvin temperature. R is a proportionality constant, remarkably the same for all gases, which relates the 4 variables. Equation is of simple form, which belies the effort expended in developing it. Over 200 years elapsed between the first quantitative experiments on gases and the final enunciation of equation. The simple form of the equation hides a remarkable sophistication in understanding. We now examine what the equation says regarding gas behaviour.

Volume-Pressure Relation

If we rearrange the equation to solve for the volume, the equation states that volume decreases as the pressure exerted on (also by) the gas increases. The inverse relation of pressure and volume is a fact of every day experience.

For example, when you compress the air in the chamber of a tire pump, you feel the increase in pressure of the air within.

In 1662, Robert Boyle developed the quantitative relationship between pressure and volume that we now know as Boyle's Law, which states that the volume occupied by a fixed amount of gas at a fixed temperature varies inversely with the pressure exerted on the gas. Boyle's Law was the first-discovered component of equation. Note that the ideal gas law predicts that the volume should fall to 0 as pressure gets very large. For real gases this does not happen.

At ordinary temperatures, most real gases will liquefy if pressure is made large enough. Once liquefication has occured, volume does not decrease much with increasing pressure. Real gases obey equation closely only when P is not too large. This is why equation is called the ideal gas law — it applies strictly only to an idealized gas, which does not liquefy but instead shrinks to zero volume when pressure is made very large.

Volume-Temperature Relation

The ideal gas law states that the volume occupied by an amount of gas, n, varies directly with the absolute temperature of the gas. A particular sample of gas, then, should double in volume if heated from 300 to 600 K. This behaviour, too, is quite well known, at least qualitatively, in every day experience, as in the expansion (contraction) of a heated (cooled) balloon.

The relationship between gas volume and Celsius temperature was first established by Jacques Charles in 1800. He clearly established the linear relationship between the two quantities. Similar experiments relating gas pressure and temperature were carried out at about the same time by Gay-Lussac, and similar proportionality was found. In both cases, the relationships obtained were not so simple as the ideal gas law, because the absolute scale of temperature was not then recognized. The relation between volume and Celsius temperature is in equation.

$$V = a \times t(^{\circ}C) + b$$

Plots of V versus t(°C) for gas samples containing different amounts of gas give different slopes and different intercepts on the V axis.

However, they all give the same intercept on the t axis, -273.16 °C. This suggests that the zero of temperature be shifted 273.16 Celsius degrees to the left. The resulting scale of temperature is, of course, the Kelvin scale. Equation takes the much simpler form, $V = a \times T$.

A profound implication of equation is that there exists a temperature at which the volume of gas should become zero. This temperature is called absolute zero. The Kelvin scale is set up so that 0 K corresponds to the temperature at which the volume of an ideal gas sample would become zero.

This corresponds to –273.16 on the Celsius scale. Any temperature lower than this value would cause the gas to have negative volume. Since this is physically meaningless, the conclusion is that temperatures lower than absolute zero do not exist. That there is a lower limit on attainable temperature is not obvious from our everyday experience and from the manner in which we normally measure and think about temperature; the conclusion is nonetheless correct.

Relation Between Volume and Amount of Gas

The ideal gas law states that the volume occupied should

increase in direct proportion with the amount (moles) of gas, when the gas is kept at the same pressure and temperature throughout.

This is an extension of the hypothesis by Amadeo Avogadro in 1811 that equal volumes of different gases contain the same number of molecules. Avogadro made this statement in explanation of many observations about the relative volumes of gaseous elements that react to form compounds. It was in fact this hypothesis that finally allowed correct molecular formulas to be deduced and the atomic mass scale to be put on a firm basis.

The Value of the Gas Constant R

The gas constant R turns out to be an ubiquitous and extremely important quantity in physical science. We begin by obtaining a value for it based on the experimentally-measured volume of 1.0 mole of gas at 1.00 atm pressure and a temperature of 273.16 K.

It is found that under these conditions the gas occupies a volume of 22.414 L. Rearranging equation to solve for R, we can calculate its value:

$$R = PV/nT = (1.000 \text{ atm})(22.414 \text{ L})/ (1.000\text{mole})(273.16\text{K}) = 0.08206 \text{ L-atm/K-mole}$$

The numerical value of R depends upon the units in which pressure and volume are expressed. P is commonly expressed in atmospheres or torr (both non-SI) or Pascal (SI), and V is commonly expressed in mL. Values of R in common P-V unit combinations are presented in Table.

Table. In Common Pressure-Volume Unit Systems

Pressure Unit	*Volume Unit*	*R*
Atm	L	0.08206 L-atm/K-mole
Bar	L	0.08314 L – bar/K-mole
Torr	mL	6.237×10^4mL-torr/K-mole
Pa	m^3	8.314 J/K-mole

From the Pascal-m^3 value, we see that R has units of energy/K-mole. It follows that the combined units L-bar, L-atm, and mL-torr are also energy units. We can express R in terms of any desired energy unit by using an appropriate conversion factor. In using the ideal gas law, equation, the units of P, V, and R must be self-consistent, and the temperature must be expressed in Kelvins. A few examples of the use of the ideal gas law are now presented.

Example: An ideal gas occupies 326 mL at 730 torr of pressure and a temperature of 19°C. What is its volume at 1.00 atm and 0°C?

Example: We take two approaches to the solution of this problem. In the first approach, we use the ideal gas law twice. First, we calculate the moles of gas under the original V,T,P conditions; second, we calculate the volume occupied by this number of moles at 1 atm and 0°C.Convert to the necessary units of T, P:

Initial	*Final*
T(K) = 273 + 19 = 292 K	T(K) = 273 K
P = 730 torr = 730/760 atm = 0.961 atm	P = 1.00 atm
V = 326 mL = 0.326 L	V is unknown

Calculate the number of moles of gas under the initial conditions:

n = PV/RT

= (0.961 atm)(0.326 L)/(0.08206 L-atm/K-mole)(292K)

= 1.307×10^{-2} moles

Calculate volume at final T and P:

V = nRT/P

= (1.307×10^{-2} moles)(0.08206

L-atm/K-mole)(273K)/ (1.00 atm)

= 0.29 L (2 significant figures)

To shortcut this long process we take advantage of the constancy of the amount (moles) of gas. Since n is constant, PV/RT must also be constant, and its initial and final values are the same:

$$P_iV_i/T_i = P_fV_f/T_f$$

Solving for V_f and substituting numbers gives V_f = 0.29 L.

Example: 3.24×10^{-4} moles of ideal gas occupies 5.31 mL at 25°C. Calculate the gas pressure.

Example: Convert units, and apply the ideal gas law to the problem.

Convert units:T(K) = t(°C) + 273

= 25 + 273

= 298KV(L)

= 5.31 mL/(1000 mL/L)

= 5.31×10^{-3}

L*Calculate P*:P = nRT/V

= $(3.24 \times 10^{-4})(0.08206)(298)/(0.00531)$

= 1.49 atm

Example: Consider the apparatus in Figure. Initially, a quantity of gas is confined in the left box at a pressure of 3.40 atm and a temperature of 286 K. 4 sequential processes, a, b, c, and d, are carried out. Work out the value of each indicated quantity after the completion of each process.

- Valve 1 is opened with valve 2 closed. T is kept constant. What are the final volume and pressure of the gas?
- T is raised to 350 K. What are the final V and P of the gas?
- Valve 1 is closed, valve 2 is opened. What are the final V and P of the gas to the right of valve 1?
- T is lowered to 300 K. What are the final V and P of the gas to the right of valve 1?

Solution: We apply the ideal gas law to each process in turn.

- When valve 1 is opened, the gas will expand to occupy its original box and the 3.0-L box at uniform pressure. T is constant, and the amount of gas is too. The gas volume goes up by a factor of 4, so the pressure goes down by the same factor.
  V occupied by gas = 4.0 LP of gas = 3.4/4 = 0.85 atm
- When T is raised to 350 K, the gas pressure will increase in direct proportion, because V will remain constant. Therefore
  V occupied by gas = 4.0 LP of gas = T_fP_i/T_i = (350)(0.85)/(286) = 1.04 atm

- When valve 1 is closed, the gas in the 1.0-L box will no longer communicate with the gas in the middle box. Only the gas in the middle box will be involved when valve 2 is opened. This gas will expand from its initial pressure of 1.04 atm until its final pressure is 0.25 atm. At this point, the piston in the cylinder will stop moving. V occupied by gas to right of valve 1 = P_iV_i/P_f = (1.04)(3.0)/(0.25) = 12.48 L.P of gas to the right of valve 1 = 0.25 atm
- When T is lowered to 300 K, the volume occupied by the gas will decrease, as the piston moves in to maintain pressure at 0.25 atm.
  V occupied by gas to the right of valve 1 = V_iT_f/T_i = (12.48 L)(300)/(350) = 10.70 L.P of gas to the right of valve 1 = 0.25 atm

Example: 0.432 g of an ideal gas exerts pressure of 0.46 atm in a volume of 768 mL at 0°C. Calculate the molar mass of the gas.

Solution: The ideal gas law can be expressed directly in terms of MM because moles is mass over MM:

n = PV/RT = (mass)/MM

Rearrange for MM:

MM= mRT/PV, where m is mass.

Solve:

MM =(0.432g)(0.08206 L-atm/K-mole)(273 K)/(0.46 atm)(0.768 L) = 27.4 g/mole

The gas could be nitrogen, N_2; carbon monoxide, CO; or ethylene, C_2H_4.This method for determining the MM of an unknown gas or volatile liquid is called the Dumas method. It is simple, fast, and reasonable accurate.

Example: Calculate the density of oxygen gas at a pressure of 0.940 atm at room temperature (25°C).

Solution: Rearrangement of the ideal gas law with the recognition that density is mass per volume gives density directly.

Rearrange: PV = nRT

n/V = m/(MM)V = r/MM = P/RT

$\rho = P(MM)/RT$

Solve the equation with the specific variables given:

$= (0.940 atm)(32.00 g/mole)/(0.08206 L\text{-}atm/K\text{-}mole)(298 K)$

$= 1.23 g/L$

The value has units g/L, and is about 1000 times less than the density of water at the same temperature. Gases are much less dense than liquids, implying that the gas molecules are much further apart than are liquid molecules.

Dalton's Law of Partial Pressure

The ideal gas law can be applied to mixtures of gases as well as to single pure gases. Consider a mixture of two non-reacting gases, A and B, in a container of volume V. According to the ideal gas law, the pressure exerted by the mixture is given by equation.

$$P_T = n_T RT/V$$

Here n_T, the total moles of gas, is the sum of the moles of A and the moles of B:

$$n_T = n_A + n_B$$

Substitute equation and expand:

$$P_T = (n_A + n_B)RT/V = n_A RT/V + n_B RT/V$$

Suppose that we could remove all of the molecules of B from the container. What pressure would A exert in the container alone? This is easily obtained from the ideal gas law to be $P_A = n_A RT/V$. Similarly, $P_B = n_B RT/V$. Substituting the expressions for P_A and P_B into equation gives:

$$P_T = P_A + P_B$$

This is Dalton's Law of partial pressures. In words, it says that the total pressure of a mixture of gases (P_T) is the sum of the partial pressures of the gases composing the mixture. The partial pressure of gas A is the pressure that n_A moles of A would exert if present alone in the container.

Dividing the expression for P_A by equation gives:

$$P_A/P_T = n_A/n_T = n_A/(n_A + n_B)$$

The ratio of the moles of A to the total moles of all gases

in the system is called the mole fraction of A, and is symbolized X_A. Thus $X_A = n_A/n_T$. Substituting this into 5-2-7 and rearranging, we obtain equation:

$$P_A = X_A P_T$$

The partial pressure of A in a mixture of A and other gases is the total pressure multiplied by the mole fraction of A.

Example: A gas mixture containing equal numbers of nitrogen and oxygen molecules exerts 1.32 atm pressure. What is the partial pressure of nitrogen?

Solution: Since half the molecules are nitrogen, half the moles are also nitrogen, so the mole fraction of nitrogen is 0.5.

$$P_{N2} = 0.5\ (1.32\ \text{atm}) = 0.66\ \text{atm}$$

We close this section by emphasizing an implicit assumption of the above treatment. Figure shows two gases, A and B, in a box. If the box has volume V, what fraction of this volume is occupied by A, and what fraction by B? A frequent answer might be that since half the molecules in the box are A, A occupies half the volume.

This is, however, incorrect. If, when we fill the box with gas, we put A in first, the molecules of A will spread out to occupy the whole box. Now add some molecules of B.

Since these molecules are "unaware" of the presence of A, they, too, will spread out in the entire volume. Thus both gases occupy the same volume — the entire volume of the container. $V_A = V_B = V$, the box volume. This is implicit in equations. Similarly, we assumed that temperature is the same for both gases:

$$T_A = T_B = T.$$

In contrast, the total moles, n_T, is the sum of the moles of A and B; and the total pressure is the sum of the partial pressures. These relationships are summarized in Table.

Table. Relationships Among State Variables for a Gas Mixture

$T_A = T_B$	$= T_{mixture}$
$n_A + n_B$	$= n_{mixture}$
$P_A + P_B$	$= P_{mixture}$
$V_A = V_B$	$= V_{mixture}$

Dalton's Law is particularly useful in experimental situations involving the collection of gases over liquids. In such situations, the space above the liquid contains not only molecules of the gas, but also molecules of the vapour form of the liquid, produced by evaporation of the liquid. The vapour exerts a partial pressure that is part of the total pressure exerted by the gas mixture over the liquid.

This partial pressure due to vapour that exists above the corresponding liquid phase is the vapour pressure, P_{vap}, of the liquid. According to Dalton's Law,

$$P_T = P_{gas} + (P_{vap})\text{liquid}$$

KINETIC MOLECULAR THEORY

The ideal gas law was developed from macroscopic observations, with no knowledge of the behaviour of a gas at the molecular level. We now attempt to interpret the ideal gas law in terms of a reasonable molecular model (simple physical picture) of gas behaviour.

What are the individual molecules doing, and why does their behaviour manifest itself in the ideal gas law? We now enter the domain of theory – a mental interpretation of experimental results.

The model of gas behaviour currently accepted by scientists is the Kinetic Molecular model. The theory of gas behaviour built on this model is called the Kinetic Molecular Theory (KMT). This theory is one of the oldest and most resoundingly successful in science.

First, we assume that gases consist of very small molecules. Since a gas occupies the entire volume of its container, and since gases readily diffuse, the molecules must move about in space. Consequently they have kinetic energy — energy of motion.

Since gases are compressible, the molecules must be far apart. Focussing on an individual molecule, it seems reasonable to assume that it is a small particle that moves about at random, occasionally colliding with a wall of the container and with other gas molecules. The molecule sometimes gains and sometimes loses energy in these collisions, so that it frequently changes speed. Sometimes it moves rapidly, sometimes slowly. Thus our picture of the gas is dynamic rather than static. Figure is an attempt to portray this simple physical model of the gas phase.

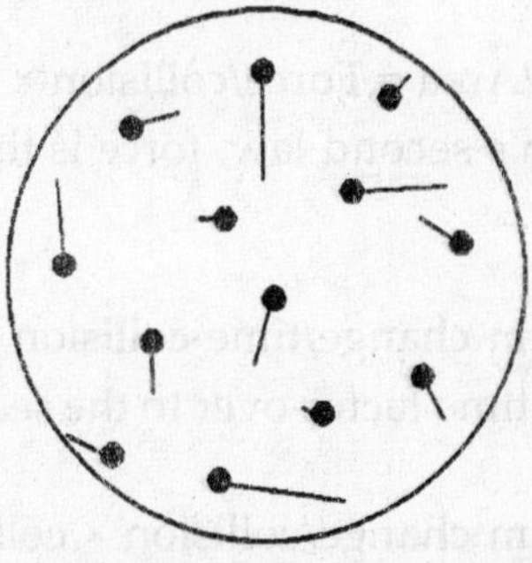

Fig. Dynamic Model of the Gas Phase

These statements about gas molecules are consistent with the properties of gases listed. None of them have been proven, but they seem reasonable and constitute the postulates of the kinetic molecular theory:

- Gases consist of molecules that behave like tiny hard spheres;
- The molecules have no volume — they may be treated as points;
- There are no forces of attraction between molecules;
- Collisions between molecules are elastic (kinetic energy is conserved);
- The molecules are in ceaseless random motion, frequently colliding with the container walls and with each other. A collection of gas molecules is highly disordered.
- The average kinetic energy of the molecules is proportional to Kelvin temperature.

We now pursue the consequences of these postulates. Postulate 6, the origin of which is not obvious. Our aim is to develop an expression for pressure.

At the molecular level, pressure must result from collisions of gas molecules with the container walls. A molecule hitting the wall exerts a force on it.

The collection of all such forces on a unit area of wall during a given time interval constitutes the pressure. We should be able to calculate P from the force exerted by each collision, multiplied by the number of collisions per unit area of wall:

P = Force/Area = Force/collision × Collisions/area

From Newton's second law, force is the rate of change of momentum.

Thus,

F = momentum change/time-collision × collisions/area

We move the time factor over to the second term to obtain equation.

F = momentum change/collision × collisions/time-area

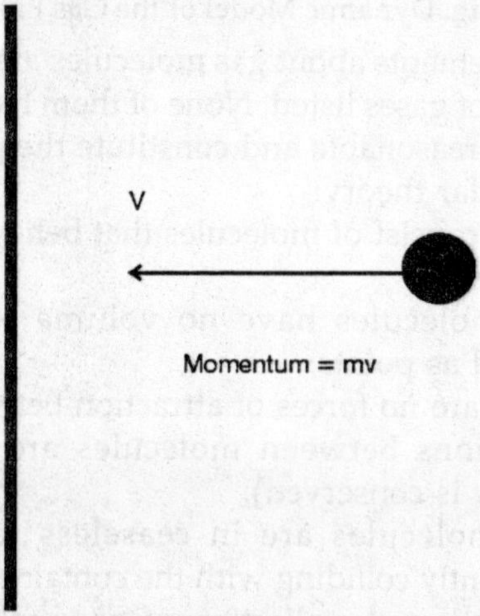

Fig. Collision of Gas Molecule with Container Wall

To obtain the momentum change per collision, envision a molecule moving directly toward a wall with velocity v, as shown in Figure. Since momentum is conserved in any collision, the change in momentum of the molecule is,

momentum change = momentum after collision– momentum before= m(-v)–m(v) = –2mv

The change of momentum of the wall must therefore be

2mv to give a total change of zero. We now have the first term in equation:

$$\text{momentum change/collision} = 2mv$$

We take an intuitive approach to the second term. The number of collisions per unit time per unit area of wall should depend on 1) the number of molecules per unit volume in the container, N/V (the more there are, the more collisions there should be with the walls); and 2) the speed v at which a molecule moves (the faster the movement, the more collisions per unit time that should occur). Equation is based on these ideas:

$$\text{collisions/time-area} = (N/V)(v)$$

If we assume for simplicity that the container is cubical, there are 6 walls over which these collisions must be spread. *The collisions with a particular wall are then*:

$$\text{Collisions/time-area} = Nv/6V$$

Despite our restrictive assumptions about the shape of the container, equation turns out to be valid for a container of any shape!. We now multiply the expressions in equations to obtain the pressure:

$$P = (2mv)(Nv/6V) = Nmv^2/3V$$

Rearrangment gives equation:

$$PV = Nmv^2/3$$

This is as far as our molecular model takes us.

We are now at the interface between theory and experiment. Experiment (the ideal gas law) relates pressure and volume to temperature:

$$PV = nRT$$

Theory, equation, relates pressure and volume to the mass and speed of a gas molecule. To bring experiment and theory into correspondence, we must equate the right sides of the preceding two equations.

In order for KMT to successfully explain the ideal gas law, it is necessary that:

$$nRT = Nmv^2/3$$

Equation can be simplified by replacing n with N/N_o (N_o

is Avogadro's Number); $mv^2/2$ by KE (the kinetic energy of an average gas molecule); and solving for KE. The result is

$$KE_{molecule} = 3RT/2N_o$$

We refine this equation by making two further realizations. First, since R and N_o are both constants of nature, so must their ratio be. It is symbolized k, and is called Boltzmann's constant. Its value is 1.381×10^{-23} J/K-particle.

Second, we have recognized that the kinetic energy of a gas molecule changes frequently as it undergoes collisions. However, its average speed over time is constant. The refined equation is among the most important equations in science:

$$\text{Average } KE_{molecule} = 3kT/2$$

The average kinetic energy of a gas molecule, no matter what its chemical identity, depends only on the Kelvin temperature.

This gives a deep insight into the concept of temperature: it is a measure of the average kinetic energies of the molecules of a substance, hence a measure of their average speed. As T is increased, molecules move faster; as it is lowered, they move slower.

The temperature at which molecular motion ceases is absolute zerc. Temperatures lower than absolute zero are impossible because a molecule may not have negative kinetic energy.

Equation is simple and profound. Is there an experiment that can be done to test its validity? There is in fact a simple experiment that verifies equation in a rearranged form.

If we replace the average kinetic energy of a molecule with the expression $m(v^2)_{avg}/2$, where $(v^2)_{avg}$ is the average of the square of the speed, and solve the resulting expression for $(v^2)_{avg}$, we obtain equation.

$$(v^2)_{avg} = 3kT/m$$

The square root of $(v^2)_{avg}$ is called the root mean square speed, and is for our purposes approximately equal to the average molecular speed. Making this equality gives equation.

$$v_{avg} = (3kT/m)^{1/2}$$

Multiplying both numerator and denominator of the

argument on the right side of this equation by Avogadro's number N_o gives the useful variant in equation.

$$v_{avg} = (3RT/MM)^{1/2}$$

The implication of equations is that a heavy gas molecule moves more slowly than a light one, in a quantifiable way. The ratio of the speeds of the light (L) and heavy (H) molecules should be the square root of the inverse ratio of their molar masses:

$$(v_H/v_L)_{avg} = (MM_L/MM_H)^{1/2}$$

Thomas Graham measured the rates of diffusion of various gases. Diffusion is the process by which gases disperse in space via random molecular motion. The results of Graham's experiments are summarized in equation rate diffusion of gas A/rate of diffusion of gas B =(density of gas B/ density of gas A)$^{1/2}$.

But we have previously seen that gas density is proportional to MM. If we make the logical assumption that a gas diffuses at a rate that is directly proportional to the average speed of its molecules, then Graham's Law of Diffusion is in exact agreement with equation.

Example: Calculate the ratio of the speeds of H_2 and CO_2 molecules at 25°C.

Solution: As long as temperature is the same for both molecules, its actual value is unimportant.

$$v_{H2}/v_{CO2} = (MM_{CO2}/MM_{H2})^{1/2} = (44.0/2.02)^{1/2} = 4.67$$

Example: Calculate the average speed of a molecule of nitrogen gas at room temperature, 25°C.

Solution: $v_{N2} = (3RT/MM)^{1/2}$

$= (3(8.314\ \text{J/K-mole})(298\ \text{K})/(0.02801\ \text{kg/mole}))^{1/2}$

$= 5.15 \times 10^2\ \text{m/s}$

This is 1150 miles per hour, or somewhere between the speeds of a commercial jet liner and a fighter jet. Please note that in applying equation in Example, values of R and MM with appropriate units have been used, so that speed comes out with appropriate units.

Use of R in units of, for example, atm-L, produces a speed with absurd units. The curve is a plot of the fraction of molecules having speed s, f(s) (we use s for speed, v for velocity), versus the possible values of speed between zero and infinity.

The curve has several noticeable features:

- *The curve has the expected shape*: Low on both ends, indicating that few molecules have extremes of speed; and peaking in the middle, indicating that most molecules have speeds somewhere between the extremes.
- The curve is unsymmetrical because there is a definite lower limit (zero) but no definite upper limit to speed.
- The total area under the curve is the total number of gas molecules.
- The height of the curve at a particular speed (say s_1) is proportional to the fraction of molecules having speed s_1.
- The area under the curve between two speeds s_1 and s_2 is the fraction of molecules haveing speeds in the range s_1 to s_2.
- The mathematical expression for f(s) obtained by *Maxwell and Boltzmann is equation*:

$$f(s) = (\text{constant})(s^2) \exp(-ms^2/2kT)$$

- The s^2 term increases with increasing s and the exponential term decreases with increasing s. Thus f(s) first increases, but then peaks and decreases with increasing s as the exponential term takes over. The speed at the maximum of the curve, s_{mp}, is the most probable speed, because the greatest fraction of molecules have it.

The Maxwell-Boltzmann Distribution Law. We have said that in a gas the molecules move randomly, and that the speed of a molecule changes frequently.

In the mid 1800's, James Clerk Maxwell in England, and Ludwig Boltzmann in Austria, were concerned with using statistical methods to describe, precisely and mathematically, the distribution of molecular speeds.

The result of their efforts is called the Maxwell- Boltzmann Distribution Law. They showed that at a given temperature, the molecular speed distribution follows a curve like that in Figure.

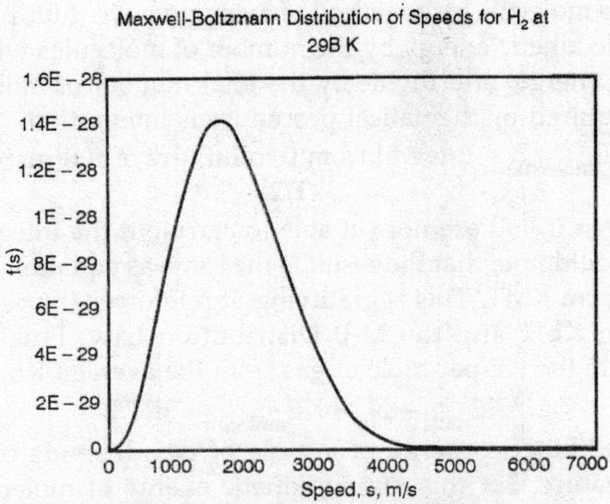

Fig. Maxwell-Boltzmann Distribution

- As T increases, the curve flattens and the maximum moves to higher speed, because the speeds of all molecules tend to increase with increasing T. The area under the curve stays constant, however, as long as the number of gas molecules is unchanged.

We now do two calculations using equation. First, we calculate the most probable speed, s_{mp}. This is a good approximation to the average speed, but the average will be somewhat larger because the curve is biased to higher speeds. s_{mp} is the speed at which f(s) is maximum..

At this maximum, the slope of the plot — the derivative of f(s) — is zero:

$$df(s)/ds = d/ds\ [(s^2)(\exp(-ms^2/2kT))] = 0$$

$$= ds^2/ds\ \exp(-ms^2/2kT) + s^2\ d[\exp(-ms^2/2kT)]/ds = 0$$

$$= 2s\ \exp(-ms^2/2kT) - s^2(ms/kT)\exp(-ms^2/2kT) = 0$$

Dividing by the exponential, we obtain:

$$2s - ms^3/kT = 0$$

This is readily solved for s_{mp}:

$$s_{mp} = (2kT/m)^{1/2}$$

As expected, this is a bit smaller than the root-mean-square speed in equation. Next we calculate the average KE of a gas molecule by a method of averaging; we multiply each possible kinetic energy by the number of molecules with that kinetic energy, and divide by the total number of molecules. The required mathematical procedure is integration.

$KE_{molecule}$ = Integral from 0 to Infinity of $f(s)(ms^2/2)ds$ = $3kT/2$

Even if you are not yet able to carry out the integration, you should note that the result is the same as equation, which came from KMT. This is gratifying. It reinforces the validities of both KMT and the M-B Distribution Law. Finally, we calculate the KE per mole of gas from the average $KE_{molecule}$:

$$KE_{mole} = N_o \times KE_{molecule} = 3RT/2$$

The kinetic energy of a mole of gas depends only on temperature. For this reason, kinetic energy of molecules is called thermal energy. Summary of the Interpretation of the Ideal Gas Law in Terms of Molecular Behaviour. The relationship between the macroscopic and microscopic views of gas behaviour can be summarized in several statements.

- Gases are compressible because they consist of small molecules that are far apart.
- Molecules of a gas are in constant motion, allowing gases to diffuse and to flow.
- Pressure results from collisions of gas molecules with the container walls.
- Pressure increases as volume decreases (Boyle's Law). A decrease in V increases the number of molecules per unit volume, leading to more molecular collisions with a unit area of the wall per unit time.
- An increase of temperature causes an increase in pressure (the Law of Gay-Lussac). KE of the molecules increases with T. Faster molecules cause an increase in pressure for two reasons:
  - There are more collisions with the wall per unit time;

  - Each collision exerts a greater force, since the molecule is moving faster.
- An increase in the amount (moles) of gas causes an increase in pressure at constant T (Avogadro's hypothesis). More gas means larger N/V. Higher N/V means more collisions per unit time and higher pressure.
- The ratio of diffusion rates of two gases is proportional to the square root of the inverse ratio of their molar masses (Graham's Law). This is a consequence of the average kinetic energy of a gas molecule being dependent only on T. At a particular T, a heavy molecule moves slower than a light one, but its larger mass offsets its smaller speed, giving the same kinetic energy. An important consequence of this is that a sample of gas with heavy molecules will exert the same pressure as a sample of gas with the same number of light molecules, as long as their temperatures and volumes are the same. Both samples satisfy the ideal gas law, independent of what their molecules are like. Although heavy molecules move slowly and strike the wall less often, they strike it with bigger mass, and cause the same pressure.

ROLE OF GAS PHASE CHEMICAL REACTIONS

The first decade of the 19th century must have been both exhilarating and frustrating for chemists. On the one hand, Dalton's atomic theory rationalized many known facts about chemical elements, compounds, and reactions. In particular, the theory provided an explanation for the laws of conservation of mass, definite proportions, and multiple proportions, and gave a rational basis for the idea of reproducible combining masses of the elements in compound formation (e.g., the combining masses of oxygen and hydrogen in forming water are in the ratio 7.94 to 1).

On the other hand, attempts to develop a scale of atomic masses were thwarted by the ignorance of chemical formulas.

Chemists, led by Dalton, assumed that the formula for water was HO, and that water was formed from hydrogen and oxygen by equation:

$$H + O \rightarrow HO$$

This was the simplest assumption to make in the absence of definite knowledge of the formulas of elemental hydrogen, elemental oxygen, and water.

In 1809, Joseph Gay-Lussac conducted a series of experiments with gases that ultimately enabled the development of the atomic mass scale. Specifically, for reactions in which gases react to form other gases, he studied the relationships among the volumes of gaseous reactants consumed and the volumes of gaseous products formed. His result is of tremendous importance.

Gay-Lussac discovered that, as long as the experiments were conducted at a constant temperature and pressure, the volumes of reactants used and products produced always gave whole-number ratios.

One of many reactions that he studied was that between hydrogen and oxygen to form water vapour. He found consistently and reproducibly that for every one volume of oxygen used, two volumes of hydrogen were required, and two volumes of water vapour were produced:

$$2 \text{ vol hydrogen} + 1 \text{ vol oxygen} \rightarrow 2 \text{ vol water vapour}$$

Gay-Lussac was not certain what to make of these results; however, Amadeo Avogadro was. He made the (to him) reasonable proposal that Gay-Lussac's results implied that equal volumes of gases, at the same temperature and pressure, contain equal numbers of molecules. In modern terms, we say that the volume occupied by a gas is proportional to the number of moles of gas present.

Operating from this assumption, he concluded and stated that Dalton's view of the hydrogen/oxygen reaction, expressed in 5-4-1, was incorrect, because it was not consistent with the two-to-one-to-two hydrogen/oxygen/water volume ratios that Gay-Lussac observed.

He then went on to propose that hydrogen and oxygen occur as diatomic molecules, and that water in fact contains two atoms of hydrogen per atom of oxygen, so that the water formation reaction becomes:

$$2H_2(g) + O_2(g) \rightarrow 2H_2O(g)$$

This is the simplest proposal consistent with the combining volumes observed; as we know today, it is entirely correct.

DOUBLE DISPLACEMENT REACTIONS

Three broad classifications of chemical reactions: electron transfer processes, proton transfer processes, and double displacement processes.

We defined a double displacement reaction as one in which the positive and negative portions of the reactant molecules interchange to form products. It is appropriate on gases to focus briefly on double displacement reactions that produce a gas as one of the products. An important series of examples of such reactions involves metal carbonate compounds (called carbonate salts) and acids as reactants. The reaction of a metal carbonate with an acid is shown generically in equation.

$$MCO_3(\text{s or aq}) + HX(aq) \rightarrow MX_2(aq) + H_2CO_3(aq)$$

This reaction does not have a gas as one of its products. However, in a subsequent step, carbonic acid, H_2CO_3, decomposes to water and carbon dioxide, a gas that escapes in part from the aqueous solution in the form of effervescence:

$$H_2CO_3(aq) \rightarrow H_2O + CO_2(g)$$

A specific example of the acid-carbonate reaction. Here the net reaction, obtained as the sum of equations, is shown.

$$Na_2CO_3 + 2\ HCl \rightarrow 2NaCl + H_2O + CO_2$$

Acid-carbonate reactions are very important in at least two contexts. First, metal carbonates that have low solubility in water make up a substantial fraction of the earth's crust (limestone is an example).

Deposits of carbonate rocks in contact with slightly acidic ground waters undergo reactions. In so doing, they serve to maintain the acidity level of the water at a fairly constant level, supporting aquatic plant and animal life.

Second, reactions are important in the context of acid rain, which results when nitric acid, HNO_3, and sulfuric acid, H_2SO_4, are produced in the atmosphere from SO_2, produced in coal combustion, and NO_2, a byproduct of the internal combustion engine. Limestone (primarily calcium carbonate) is widely used in construction, and has been used for centuries as a medium for sculptors.

Repeated and prolonged contact of building edifices and priceless works of art with acid rain results in their slow destruction as calcium carbonate reacts with nitric and sulfuric acids according to equations.

Building edifices can be replaced; works of art cannot. Finally, dietary calcium tablets are made primarily of calcium carbonate.

Contact of a stomach tablet with stomach acid (hydrochloric acid) causes the tablet to dissolve via reaction, after which the calcium can be utilized by the body.

$$CaCO_3(s) + HCl \rightarrow CaCl_2(aq) + H_2O + CO_2(g)$$

Metal sulfites (for example, Na_2SO_3) react with acids in analogous fashion to form aqueous sulfurous acid, H_2SO_3. This subsequently decomposes to water and sulfur dioxide:

$$Na_2SO_3(aq) + HNO_3(aq) \rightarrow NaNO_3(aq) + H_2SO_3(aq)$$

$$H_2SO_3(aq) \rightarrow H_2O + SO_2(g)$$

Example: A 500-mg calcium tablet dissolves in an excess of stomach acid. What volume of CO_2 is produced?

Solution: We begin with a balanced chemical equation for the reaction of stomach acid and the calcium tablet, which we assume to be pure calcium carbonate:

$$CaCO_3(s) + 2HCl(aq) \rightarrow CaCl_2(aq) + H_2O + CO_2(g)$$

To use the numerical information in the equation coefficients, we must work in moles:

$$\text{moles } CaCO_3 = 0.500 \text{ g/(100.086 g/mole)} = 5.00 \times 10^{-3} \text{ moles.}$$

1 mole of CO_2 is produced for each mole of calcium carbonate that reacts. It follows that 5.00×10^{-3} moles of CO_2 are formed. The ideal gas law enables us to calculate the volume of CO_2. We assume the temperature in the stomach to

be 98.6 °F, or 37 °C, and that the carbon dioxide is produced at 1 atm pressure.

Then,

$$V = nRT/P$$

$$= (5.00 \times 10^{-3} \text{ moles})(0.08206 \text{ atm-L/ mole-K})(310 \text{ K})/(1.00 \text{ atm})$$

$$= 0.127 \text{ L}$$

This volume of CO_2 is probably not sufficient to cause "indigestion."

Chapter 8

Stoichiometry

REACTION STOICHIOMETRY

The idea that a chemical equation for a reaction is essentially a recipe. Our purpose in doing that was to ground the rather abstract notions of chemical formulas and equations to something concrete, with which almost everyone has experience. All of the important quantitative aspects of chemical equations—collectively known as reaction stoichiometry—can be understood in terms of the recipe idea. It is instructive to return to the fruit salad recipe for our first example.

Example: You intend to make 20 fruit salads for a large dinner party according to the following recipe:

1 apple + 5 oranges + 10 grapes → 1 fruit salad

You go to the store to purchase starting materials, and come home with the 10 pounds of apples, 15 pounds of oranges, and 1 pound of grapes. Are these materials sufficient to produce 20 fruit salads?

Solution: We have a problem at the outset: the equation is expressed in terms of numbers of apples, oranges, and grapes; however, we know only the total amounts of each fruit by mass. Clearly we need more information, specifically, the average mass of an apple, an orange, and a grape. Suppose that we can weigh a typical apple, orange, and grape on a simple kitchen scale. Suppose further that the apples, oranges, and grapes are of unusually uniform size, so that the weight of one apple faithfully represents the weight of any apple. Our weighings give the following results:

1 apple-weight = 0.25 lb/apple

1 orange-weight = 0.33 lb/orange

1 grape-weight = 0.00033 lb/grape

These unit weights enable us to convert grocery store masses into numbers of apples, oranges, and grapes:

Number of apples = 10 lbs/(0.25 lb/apple) = 40 apples

Number of oranges = 10 lbs/(0.33 lb/orange)
= 30 oranges

Number of grapes = 1 lb/(0.00033 lb/grape)
= 3000 grapes

Now we are in a position to apply the recipe, which tells us that for each apple used, we must use 5 oranges and 10 grapes. To use 40 apples would require 200 oranges and 400 grapes. We have more than enough grapes to use up all of the apples, but we do not have enough oranges. We can conclude that the number of fruit salads will be limited by the number of oranges that we have.

Because we need 5 oranges per fruit salad, we can make only 6 fruit salads from 30 oranges. This will require 6 apples and 6 × 10 = 60 grapes. We will need to return to the store to get quite a few more oranges. Specifically, we will require enough for 14 more fruit salads, or a total of 14 × 5 = 70 oranges. Since each orange weighs a third of a pound, we need somewhat over 20 pounds more of oranges.

From this example we learn a few important ideas. First, the recipe is expressed in numbers of units of each ingredient, not in terms of their masses. Similarly a chemical equation is expressed in terms of numbers of atoms, molecules, or formula units, not in terms of their masses. Second, the amounts of our fruit salad ingredients are expressed as masses when we buy them at the store. Similarly, when we buy chemical substances, the amounts are expressed as masses. Third, to apply the recipe, we must convert the masses to numbers of things using the mass per thing.

Having done this, we figure out how many fruit salads it is possible to make with the given amounts of ingredients. We must do the same thing with a chemical equation. Finally, one of the ingredients is found to limit the number of fruit salads that can be made. We will find that in many cases, one of the reactants limits the amount of product that can be formed in a chemical reaction. This will be the reactant that is used up first; it is called the limiting reactant. Let's now see how the recipe analogy helps us in doing stoichiometric calculations for chemical reactions.

Example: What mass of pure silicon can be obtained by reduction of $SiCl_4$ with 10.0 g of magnesium metal at elevated temperature? (This reaction is used in the process of extracting silicon from naturally occurring ores. Silicon is used to make semiconductors.) Assume that a large supply of $SiCl_4$ is on hand.

Solution: We require first a balanced chemical equation (recipe), which follows from the description of the process. We are told that silicon is produced by reaction of Mg and $SiCl_4$. We put Mg and $SiCl_4$ as reactants, and Si as a product. The reaction is carried out at elevated temperature, so magnesium is present as a liquid.

$$SiCl_4(l) + Mg(l) \rightarrow Si(s)$$

The equation is not and indeed can not be balanced as it is, because there are no magnesium and chlorine atoms in the products. We must first propose at least one other product for the reaction in order to balance it. The simplest choice is magnesium chloride, which can be shown experimentally to be correct.

Thus,

$$SiCl_4(l) + Mg(l) \rightarrow Si(s) + MgCl_2(s)$$

Next we must balance the equation to conform with the law of mass conservation. This is accomplished by putting coefficient 2 before $MgCl_2$ to balance chlorine; then coefficient 2 before Mg. Balancing is simple in this case.

$$SiCl_4(l) + 2Mg(l) \rightarrow Si(s) + 2MgCl_2(s)$$

At this point, we are ready to approach the problem.

The equation gives us relationships between the numbers of moles of reactants and products. Keeping in mind the recipe analogy, the following roadmap results:

- Convert the given mass of Mg to moles Mg.
- Use the coefficients in the equation to relate moles Mg (reactant) to moles Si (product).
- Convert moles Si to grams of Si.

Mass Mg → moles Mg → moles Si → mass Si.

With the exception of the last step, this is exactly what we did in the recipe calculation in Example.

Moles Mg = 10.0 g Mg/(24.305 g Mg/mole)

= 0.4114 moles Mg

Moles Si = 0.4114 moles Mg × (1 mole Si/2 moles Mg)
= 0.2057 moles Si

The stoichiometric factor 1 mole of Si per 2 moles of Mg follows directly from the chemical equation, which tells us that 2 moles of magnesium is required to produce, and is therefore equivalent to, 1 mole of Si. Finally, we convert to mass of Si:

g Si = 0.2057 moles Si × (28.086 g Si/mole) = 5.78 g Si

We can therefore expect to obtain 5.78 g of pure silicon from 10 g of magnesium. Note that the amount of $SiCl_4$ does not play a role in the problem because we are told to assume that we have more than enough of it. This is equivalent to saying that Mg is the limiting reactant.

The situation can be made more realistic by specifying starting amounts for both $SiCl_4$ and Mg. This is done in the following example.

Example: How much pure silicon can be produced by reaction of 38.0 g of Mg and 26.7 g of $SiCl_4$(l)?

Solution: In order to use the chemical equation developed in the preceding example, we must determine the number of moles of each reactant:

moles Mg = 38.0 g Mg/(24.305 g/mole Mg) = 1.563 moles Mg

moles $SiCl_4$ = 26.7 g $SiCl_4$/(169.90 g/mole $SiCl_4$)
= 0.1572 moles $SiCl_4$

The chemical equation indicates that we require 2 moles of Mg for each mole of $SiCl_4$ that we desire to react; i.e., at the least, we must have twice as much Mg in moles as we have $SiCl_4$. The numbers above show that we have more than enough Mg to react with all of the $SiCl_4$ available. We conclude that $SiCl_4$ is the limiting reagent. The amount of Si product will be limited by the amount of $SiCl_4$ available.

$$\text{moles Si} = \text{moles } SiCl_4 \times (1 \text{ mole Si})/(1 \text{ mole } SiCl_4)$$

$$\text{moles Si} = 0.1572 \text{ moles}$$

Now convert back to mass:

$$\text{mass Si} = 0.1572 \text{ moles Si} \times 28.085 \text{ g/mole} = 4.41 \text{ g Si}$$

ELECTRON TRANSFER REACTIONS

Before considering additional examples of reaction stoichiometry, we will take some time to extend our knowledge of electron transfer reactions. The reaction between Mg and $SiCl_4$ used in the preceding two examples is an electron transfer reaction. Magnesium begins as an element with uncharged atoms, but finishes as an ionic compound in which it has a charge of 2+. It has lost electrons, and has thus been oxidized. Silicon begins as a component of $SiCl_4$, in which it has a charge of 4+; and ends as elemental silicon, with zero charge. Si has been reduced because it has gained electrons. For this reaction, the charge changes are fairly easy to see. In the reaction below, however, it may not be obvious whether or not electrons are transferred, and it is certainly not obvious how many have been transferred:

$$Cr_2O_7^{2-} + C_2H_6O \rightarrow Cr^{3+} + C_2H_4O_2$$

There is a simple set of rules that can be used to readily determine the charge on an atom in any compound. The charge determined according to these rules is called the oxidation state of the atom. This is generally different from the formal charge, because the rules for assigning electrons are different. Oxidation state, like formal charge, is an electron bookkeeping procedure. The oxidation state of an element in a compound can be determined using the following simple rules for oxidation state.

The Assignment of Oxidation State

- An atom of an element in its uncombined form has oxidation state zero. Thus, each oxygen atom in O_2

has oxidation state (O.S.) = 0. Each atom in a sample of iron metal, Fe(s), has O.S. = 0.

- For simple monatomic ions the oxidation state is the charge on the ion. Thus sodium is present in NaCl as Na+, and has O.S. = 1+; chlorine is present as Cl^- and has O.S. = –1.
- In compounds and polyatomic ions, oxygen has oxidation state –2. Because oxygen is the second-most electronegative element, it either gains two electrons to form the O^{2-} ion, or has the lion's share of electrons in a covalent compound. In Na_2O, the O.S. of oxygen is -2, and that of sodium is 1+. In SO_3, the oxidation state of oxygen is –2, and that of sulfur is 6+.
- There are two exceptions to this rule. The first is hydrogen peroxide, in which the two oxygen atoms are bonded together and must be considered to equally share the electrons of that bond. The O.S. of oxygen is –1 in peroxides. The other is the compound OF_2, in which the O.S. of oxygen is 2+.
- In compounds hydrogen has oxidation state 1+. The most common ion of hydrogen is the proton, H^+. With only a few exceptions, hydrogen may be considered to have O.S. = 1+ in its compounds.
- In most compounds, the oxidation states of elements not mentioned explicitly in these rules can be calculated from the overall charge on the molecule or ion, and the known oxidation states of the remaining elements.

Let's examine a few examples.

Example: Determine the oxidation state of each element in each molecule or ion:

- NaBr
- H_2SO_4
- $Na_2Cr_2O_7$
- $HClO_4$
- C_2H_6O

Solution: NaBr. This is an ionic compound. The ions have their normal charges.

Consequently,

$$O.S.(Na) = 1^+; O.S. (Br) = 1^-.$$

H_2SO_4. This is a covalent compound. We can use the rules for the oxidation states of oxygen and hydrogen to determine the oxidation state of sulfur. Each oxygen atom has an oxidation state of 2-.

The total for four oxygens is then 8-. Each of the two hydrogens has O.S. 1+, for a total of 2+. The compound is electrically neutral overall, so the oxidation state of sulfur must be such that its sum with the total oxidation states of hydrogen and oxygen is zero:

$$2 + O.S.(S) - 8 = 0$$

This gives O.S.(S) = 6+. We can use the following schematic approach to determine unknown oxidation states.

- *Write the formula*: H_2SO_4
- Under the formula, write the oxidation state of each atom for which the oxidation state is known:
  - H_2SO_4
  - 1 + 2–
- To determine the total charge of each element, multiply the per-atom oxidation state by the number of atoms, and write the results above the formula:
  - 2 + 8-.
  - H_2SO_4
  - 1+ 2–
- Determine the total charge of the element of unknown oxidation state using the known element charges and the known overall charge on the molecule or ion, and write the result above the element.
  - 2+ 6+ 8–
  - H_2SO_4
  - 1+ 2–
- Divide the total charge of the element of unknown O.S. by the number of atoms of the element, and write the result under the formula. This is the oxidation state of the element.

– 2+ 6+ 8–
– H_2SO_4
– 1+ 6+ 2–

The oxidation state of sulfur in H_2SO_4 is 6+.

Using a similar approach gives 6+ for Cr in $Na_2Cr_2O_7$ and 7+ for Cl in $HClO_4$. We work through C_2H_6O using the shorthand system developed above to obtain 2- as the oxidation state for carbon:

4 – 6 + 2 –
G H_6O
2 – 1 + 1

We are now equipped to examine reaction for changes in oxidation states. Using our newly developed methods, we conclude that the oxidation state of chromium decreases from 6+ to 3+ as a result of the reaction, while that of carbon increases from 2- to 0. The reaction is clearly electron transfer.

$$Cr_2O_7^{2-} + C_2H_6O \rightarrow Cr^{3-} + C_2H_4O_2$$

To develop a method for balancing the equations for redox reactions like that in reaction, we take advantage of the fact that in all electron transfer processes, the total number of electrons lost by the oxidized species is exactly the same as the total number of electrons gained by the reduced species. This is seen not to be true in reaction, because the two chromium atoms of $Cr_2O_7^{2-}$ gain a total of six electrons, whereas the two carbon atoms of C_2H_6O lose a total of only four. We have some balancing to do.

Balancing Electron Transfer Equations by the Method of Half Reactions

The method of half reactions is a systematic approach to the balancing of electron transfer reactions. We treat it as a series of steps.

Step: Divide the reaction into two half-reactions, one involving the oxidized element, the other involving the reduced element:

$$Cr_2O_7^{2-} \rightarrow Cr^{3+}$$

$$C_2H_6O \rightarrow C_2H_4O_2$$

Step: Balance each half reaction separately, first balancing atoms of the element that loses or gains electrons; then balancing oxygen and hydrogen; and finally, balancing electrical charge.

- *Redox element balance*: In the first half reaction, the element undergoing electron gain is chromium. To balance chromium, put coefficient 2 before Cr^{3+}. The second half reaction is already balanced in carbon.

$$Cr_2O_7^{2-} \rightarrow 2Cr^{3+}$$

$$C_2H_6O \rightarrow C_2H_4O_2$$

- *Oxygen/hydrogen balance*: Most redox reactions are carried out in aqueous solution, often in acidic solution, in which there are surpluses of water molecules and protons. Half reactions can be balanced for oxygen and hydrogen by adding H_2O and H^+ as needed. The chromium half-reaction is balanced by adding, first, 7 molecules of water to the right to balance oxygen; then 14 protons to the left to balance hydrogen. The carbon half reaction is balanced by adding one water molecule to the left, and 4 protons to the right.

$$Cr_2O_7^{2-} + 14H^+ \rightarrow 2Cr^{3+} + 7H_2O$$

$$C_2H_6O + H_2O \rightarrow C_2H_4O_2 + 4H^+$$

- *Charge balance*: Each half reaction is now balanced for charge by adding electrons to one side or the other so that the total charge on both sides of the half reaction is the same. For the chromium reaction, total charge after completion of step b is 12+ on the left, and 6+ on the right. Six electrons, each with a charge of 1-, are added to the left side to equalize charge. Similarly in the carbon half reaction, four electrons must be added to the right:

$$Cr_2O_7^{2-} + 14H^+ + 6e \rightarrow 2Cr^{3+} + 7H_2O$$

$$C_2H_6O + H_2O \rightarrow C_2H_4O_2 + 4H^+ + 4e$$

Step: Equalize electrons in the two half reactions. It is at this point that we make use of the key fact about redox reactions: that every electron produced in oxidation must be consumed in reduction. We equalize the electrons transferred if we multiply the chromium half reaction by two, and the carbon half reaction by 3:

$$2 \times (Cr_2O_7^{2-} + 14H^+ + 6e \rightarrow 2Cr^{3+} + 7H_2O)$$

$$3 \times (C_2H_6O + H_2O \rightarrow C_2H_4O_2 + 4H^+ + 4e)$$

Step: Recombine the balanced half reactions to give a balanced overall electron transfer reaction.

$$2Cr_2O_7^{2-} + 28H^+ + 3C_2H_6O + 3H_2O \rightarrow 4\ Cr^{3+} + 14H_2O + 3C_2H_4O_2 + 12H^+$$

Step: Simplify and check. Here we cancel common species, reduce coefficients to lowest whole numbers, and check to make sure that the same numbers of atoms of all types, and charges, occur on both sides. We obtain,

$$2Cr_2O_7^{2-} + 16H^+ + 3C_2H_6O \rightarrow 4\ Cr^{3+} + 11H_2O + 3C_2H_4O_2$$

This checks for balance, with 4 Cr atoms, 17 O atoms, 34 H atoms, and 12+ charge on each side.

Example: What mass of acetic acid, $C_2H_4O_2$, is obtained from reaction of 146.2 g of potassium dichromate, $K_2Cr_2O_7$ with 84.6 g of ethyl alcohol, C_2H_6O? Discuss changes in molecular stereochemistry when C_2H_6O is converted to $C_2H_4O_2$.

Solution: The balanced equation is, First, we calculate moles of each reactant:

$$\text{moles } K_2Cr_2O_7 = 146.2\ g/(294.181\ g/mole) = 0.497\ moles$$

$$\text{moles } C_2H_6O = 84.6\ g/(46.069\ g/mole) = 1.84\ moles$$

The equation tells us that we need 3 moles of ethyl alcohol for each 2 moles of dichromate anion reacted. Because 1 mole of potassium dichromate contains one mole of dichromate anion, moles $Cr_2O_7^{2-}$ = 0.497.

Using the 3 to 2 ratio moles C_2H_6O = 0.497 moles $Cr_2O_7^{2-}$ × (3 moles ethanol/2 moles dichromate) = 0.746 moles.

This is all we need to use up all of the dichromate. Clearly we have available much more ethanol than we need. Dichromate is the limiting reactant and determines the quantity of acetic acid obtained:

moles $C_2H_4O_2$ = moles $Cr_2O_7^{2-}$ × (3 moles acetic acid/2 moles dichromate) = 0.746 moles

mass $C_2H_4O_2$ = 0.746 moles × 60.052 g/mole = 44.8 g of acetic acid

The NMR spectrum of ethanol shows an upfield triplet (intensity 3), a midfield quartet (intensity 2) and a midfield singlet (intensity 1), consistent with the structural formula.

The spectrum for acetic acid shows an upfield singlet of intensity 3, and a far downfield singlet of intensity 1, consistent with the structural formula.

In ethanol, both carbon atoms are surrounded by four electron groups, tetrahedrally distributed, and each group is used to attach an atom or group of atoms to carbon.

The stereochemistry is therefore tetrahedral at both carbon atoms. The oxygen atom also has four electron groups distributed tetrahedrally, but only two are used to attach atoms. The stereochemisty at oxygen is bent.

The CH_3 carbon of acetic acid has the same stereochemistry as the CH_3 carbon of ethanol: tetrahedral. However, the second carbon atom is surrounded by only three electron groups, distributed in a trigonal planar arrangement. Since all three groups are used to attach atoms, the shape at this carbon is also trigonal planar. The stereochemistry at the singly bonded oxygen is bent.

ASPECTS OF REACTION STOICHIOMETRY

It is usually helpful in learning new concepts to have a visual representation of the concept.

Example.provides such a representation for a double displacement reaction.

Example: $AgNO_3$ and NaCl react in solution via double displacement according to

$$AgNO_3(aq) + NaCl(aq) \rightarrow NaNO_3(aq) + AgCl(s)$$

We react a fixed mass (1.500g) of $AgNO_3$ with varying masses of NaCl and weigh the amount of AgCl obtained as a solid in each reaction. We then plot mass AgCl obtained versus mass NaCl used.

- Why does the plot increase linearly in region 1?
- Why is the plot horizontal in region 2?
- What is the significance of the point where the region 1 and 2 lines meet (the breakpoint)?
- Suppose we were to plot mass NaCl/mass $AgNO_3$ on the *x*-axis. At what value of this quantity would the breakpoint occur?
- Suppose we were to plot moles NaCl/moles $AgNO_3$ on the *x*-axis. At what value of this quantity would the breakpoint occur?
- If we were to use KCl rather than NaCl, what would the various plots look like?

Solution:

- In this region, the amount of product goes up linearly with the amount of NaCl added, because there is sufficient $AgNO_3$ in solution to react with all of the added NaCl. Reaction runs until the limiting reagent, NaCl, is gone Region 1 defines the range in which NaCl is the limiting reagent.
- In region 2, the same amount of product is obtained no matter how much NaCl is added. Now the amount of product is determined by the fixed amount of $AgNO_3$ present in the solution. The same amount of product is always obtained because the amount of $AgNO_3$ is always the same. In region 2, $AgNO_3$ is the limiting reagent.
- The two lines intersect at the point where stoichiometrically equivalent amounts of NaCl and $AgNO_3$ are present. This gives the amount of NaCl that will exactly react with 1.500 g $AgNO_3$. When reaction is finished, both reactants will be gone. Neither reactant is limiting.
- What mass of NaCl is required to just use up 1.500 g $AgNO_3$? The two substances react in a 1:1 mole ratio. We convert 1.500 g $AgNO_3$ to moles, then calculate what mass of NaCl gives this same number of moles of NaCl:

$$\text{Moles } AgNO_3 = 1.500\text{g}/(169.872 \text{ g/mole})$$

$= 8.830 \times 10^{-3}$ moles

Mass NaCl $= 8.830 \times 10^{-3}$ moles $\times$ 58.443 g/mole

$= 0.516$ g

The breakpoint would occur at a value of 0.516/1.500 = 0.344.

- The breakpoint would occur at a mole ratio of 1, based on the stoichiometry of the chemical reaction.
- It would be necessary to add a larger mass of KCl to reach the breakpoint because it has a larger formula weight. The mass AgCl versus mass KCl plot would have a region 1 with smaller slope than region 1 for NaCl. However, the same limiting mass of AgCl would be reached. The plot of moles KCl/mole $AgNO_3$ would coincide with the mole/mole plot for NaCl, because the stoichiometry is the same for the two reactions.

The power of the mole concept is illustrated in the next example.

Example: In order to obtain copper metal, a sample of the ore, $CuFeS_2$, is taken through a process consisting of several steps, as detailed below:

$$2CuFeS_2(s) + 3O_2(g) \rightarrow 2CuS(s) + 2FeO(s) + 2SO_2(g)$$

$$2CuS(s) \rightarrow Cu_2S(l) + S(s)$$

$$Cu_2S(l) + O_2(g) \rightarrow 2Cu(l) + SO_2(g)$$

The copper is then solidified and refined by electrochemical methods.

What mass of copper can be obtained from 150.0 g of ore, $CuFeS_2$, by the above sequence of reactions, assuming each reaction runs at 100% efficiency?

Solution: It might at first seem necessary to work through the reactions one at a time to finally arrive at a mass of Cu. This is unnecessary, however. Seemingly complex problems of this type are really quite simple, if we take advantage of the conservation of mass. If we assume that all of the copper containing material produced in one step is used as input for the next, we can expect by the law of conservation of mass that all of the copper contained in the ore ends up as pure copper metal after the 3-step process. Thus moles Cu = moles Cu in $CuFeS_2$ = moles $CuFeS_2$ (since 1 mole of ore contains 1 mole of copper).

The calculation is simple:

Moles ore = 150.0 g ore/(183.51 g ore/mole ore) = 0.8174 moles

Moles Cu in ore	= 0.8174 moles
Moles Cu(s) recovered	= 0.8174 moles
g Cu(s)	= 0.8174 moles
Cu × 63.5 g Cu/mole	= 51.90 g Cu

It is interesting to note that the detailed stoichiometries of the three reactions are not used at all in solving this problem. You might want to try to classify each of the reactions as electron transfer, proton transfer, or double displacement.

Stoichiometric calculations are crucially important in the methods used by chemists to determine the relative amounts of elements in a new compound. Elemental analysis is the chemist's name for the process by which the identities and amounts of the elements constituting a particular compound are determined. For the numerous compounds containing carbon and hydrogen, combustion analysis is useful. In this process, a precisely known mass of compound is burned in an excess of oxygen.

The carbon in the compound is completely converted to carbon dioxide, CO_2, and the hydrogen to water, H_2O. The CO_2 and H_2O are collected separately and weighed. From the masses of CO_2 and H_2O produced, the masses of carbon and hydrogen in the original mass of compound can be obtained. From this information, their relative numbers of moles can be calculated. Masses of other elements present can be determined by separate methods, leading to an empirical formula.

Example: When 0.617 grams of a compound of carbon, hydrogen, and nitrogen is burned in excess oxygen, 1.716 g CO_2 and 0.3513 g H_2O are collected. What is the empirical formula of the compound?

Solution: The masses of carbon dioxide and water can be converted to moles.

From this information, the moles and consequently the masses of carbon and hydrogen in the original mass of compound can be obtained. The mass of nitrogen is then obtained by difference. Once the masses of all elements in the original mass of compound are known, conversion to the relative numbers of moles gives the formula.

Moles CO_2 = 1.716 g × (1 mole/44.009 g) = 3.899×10^{-2} moles
Moles C = moles CO_2 = 3.899×10^{-2} moles
Mass C = 3.899×10^{-2} × (12.011 g/mole) = 0.468 g C
moles H_2O = 0.3513 g × (1 mole/18.015 g) = 1.95×10^{-2} moles
Moles H = 2 × moles H_2O = 3.90×10^{-2} moles
Mass H = 3.90×10^{-2} moles × (1.008 g/mole) = 0.0393 g H

Now the mass of nitrogen in the original mass of compound can be obtained:

Mass N = 0.617 × 0.468 × 0.039 = 0.110 g
Moles N = 0.00785 moles

We now have:

Moles C = 0.03899
Moles H = 0.0390
Moles N = 0.00785

Dividing all 3 by moles N gives the formula: C_5H_5N

Example: A 32.1 mg sample of a compound containing carbon, hydrogen, and oxygen was burned in oxygen to produce 78.4 mg of CO_2 and 32.0 mg of H_2O. ^{1}H-NMR and IR data for the compound are given below:

NMR	IR
triplet, Δ = 1.05 (3H)	Strong band near 1720 cm^{-1}
singlet, Δ = 2.13 (3H)	
quartet, Δ = 2.47 (2H)	

What is the structural formula for the compound?

Solution: The combustion analysis data allow us to determine empirical formula. Then the NMR and IR data can be used to deduce the arrangement of atoms—the structural formula.

millimoles CO_2 = 78.4 mg/(44.009 mg/ m mole)
= 1.781 mmole CO_2

m mole C = mmole CO_2
= 1.781 mmole C

mg C = 1.781 mmole × 12.011 mg/mmole
= 21.4 mg

Here we have used the fact that the molar mass is the same in units of mg/mmole as in g/mole.

m mole H_2O = 32.0 mg/(18.015 mg/mmole)
= 1.776 mmole H_2O

mmole H = 2 × mmole H_2O
= 3.553 mmole H

mg H = 3.553 mmole × 1.008 mg/mmole
= 3.58 mg

Therefore, mg O = 32.1 – 21.4 – 3.58
= 7.12 mg

Convert oxygen mass to mmoles:

mmoles O = 7.12 mg/(15.999 mg/mmole)
=.445 mmole

Divide mmoles of each element by the smallest value, 0.445:

relative mmoles C = 1.781/0.445 = 4.00

relative mmoles H = 3.553/0.445 = 7.98

relative mmoles O = 0.445/0.445 = 1

The empirical formula is C_4H_8O. The infrared data suggest that a C = O group is present, so we use the NMR data to build a structure based around this group.

There are two 3-proton signals in the NMR spectrum, suggestive of CH_3 groups. One CH_3 group produces a singlet signal, so there are no protons on the atom attached to the CH_3 carbon atom. This CH_3 group is probably attached to the carbon atom of the C=O group.

So far, we have the following arrangement:

$$-C(=O)-CH_3$$

The second CH_3 signal is split to a triplet, suggesting that there are two protons on the neighboring atom. Consistent with this, the two proton signal is split to a quartet, suggesting coupling to a CH_3 group.

It appears that the structure is as shown below:

$$H_3C - CH_2 - C(=O) - CH_3$$

This compound is called methyl-ethyl ketone.

REACTIONS INVOLVING GASES

The properties of gases extensively, learning that a very simple relationship describes the interrelationship of volume, pressure, temperature, and moles of gas. Gases are frequently reactants, products, or both in chemical reactions, and often it is convenient to determine the extent of reaction by measuring the amount of gas produced or consumed. The following example illustrates this application.

Example: 17.24 g of liquid C_6H_{14} (hexane) is enclosed with 3.80 moles of O_2(g) in a cylinder fitted with a piston. The initial temperature of the mixture is 27°C,and the external (outside) pressure on the piston is 1.00 atm. The hexane and oxygen are then caused to react according to the following equation:

$$C_6H_{14}(l) + O_2(g) \rightarrow CO_2(g) + H_2O(l)$$

All of the hexane is used up. The heat produced by the reaction causes the temperature to rise to 77°C. The external pressure is maintained at 1.00 atm.

- Balance the equation
- Calculate the initial volume of reactants in the cylinder (assume that liquid hexane occupies a negligible volume.)
- Calculate the final volume of products and left-over reactants in the cylinder (assume that liquid water occupies a negligible volume.)

Solution: This problem will give us a workout on gas and stoichiometry concepts.

- The equation is balanced by placing 6 before CO_2, 7 before H_2O, and 19/2 before O_2. Fractions are then cleared via multiplication by 2.

  The result is:

  $$2C_6H_{14}(l) + 19O_2(g) \rightarrow 12CO_2(g) + 14H_2O(l)$$

- Initial volume of gaseous reactants in the cylinder. We initially have only liquid hexane and gaseous O_2 in the cylinder. We can ignore the very small volume occupied by the hexane. We have 3.80 moles O_2(g) at 27 °C and a pressure of 1.00 atm.

The ideal gas law gives us the volume:

V = nRT/P = (3.80 moles)(0.08206 L-atm/K mole) (300 K)/(1.00 atm) = 93.55 L

- Initial amounts of both reactants are specified; we must determine which is limiting. Oxygen is already expressed in moles, so all we need do is compute moles hexane:

moles hexane = 17.24 g/ (86.178 g/mole) = 0.200 moles hexane

The balanced equation indicates that 19 moles of oxygen are needed for each 2 moles of hexane. For 0.200 moles hexane,

moles O_2 = 0.200 moles hexane × (19 moles O_2/2 moles hexane) = 1.900 moles oxygen

Much more than this is available; hexane is limiting. Conclusions thus far: all hexane is used up 1.900 moles O_2 is used up; therefore 1.900 moles O_2 is left over CO_2 and H_2O are formed; volume of H_2O can be ignored.

To calculate the moles of CO_2 formed is simple:

moles CO_2 = moles hexane × (12 moles CO_2/ 2 moles hexane) = 0.200 × 6 = 1.200 moles CO_2

The total amount of gas at the end of reaction is 1.200 moles CO_2 + 1.900 moles O_2 = 3.100 moles gas. The volume can be calculated from the ideal gas law, using T = 77 + 273, and P = 1 atm:

V = nRT/P = (3.1)(0.08206)(350)/1 = 89.0 L

The piston moves slightly in during reaction.

SOLUTIONS AND CONCENTRATION

A solution is a mixture of pure substances that is homogeneous at the molecular level. That is, the molecules of the substances are intimately intermixed. Usually, one of the substances is a liquid, called the solvent. A relatively small amount of a second substance, usually a solid, is then dissolved in the liquid.

The substance that dissolves is called the solute. A solution is obtained when table sugar is dissolved in water. Water is the solvent, and sugar is the solute. Solutions are extremely important in chemistry.

Most chemical reactions are carried out by dissolving the reactants in a solvent, and some substances are sold commercially as solutions because they are unstable in pure form. The amount of solute per unit amount of solvent is called the concentration of the solution. There are several ways of expressing this, the most common of which is called the molarity.

Molarity, symbol M, is defined as the number of moles of solute per liter of solution. You will be expected to use and understand molarity in the laboratory; we introduce it here to initiate this process.

Example: A solution of K_2SO_4 (potassium sulfate) in water is prepared by the following stepwise procedure:

- 26.1743 g K_2SO_4 is transferred to a 500-mL volumetric flask.
- About 250 mL distilled water is added to the flask, which is then swirled to dissolve the salt. When the salt has completely dissolved,
- More distilled water is added until the water level coincides with the mark on the neck of the flask. This mark indicates a volume of exactly 500.0 mL.
- The solution is vigorously shaken to insure uniformity.

What is the molarity of the solution?

Solution: We calculate the moles of K_2SO_4 and divide by the total volume of solution prepared:

$$\text{moles } K_2SO_4 = 26.1743 \text{ g}/(174.253 \text{ g/mole}) = 0.1502 \text{ moles}$$

$$\text{molarity} = M = 0.1502 \text{ moles solute}/0.5000 \text{ L solution} = 0.3004 \text{ M}$$

Molarity is a useful concept for dealing with the stoichiometry of reactions in solution. We consider two examples here, one involving double displacement and the other involving proton transfer.

Example: 34.6 mL of the 0.3004 M K_2SO_4 solution from Example is mixed with 52.3 mL of a 0.0954 M solution of $BaCl_2$. How much $BaSO_4(s)$ forms?

Solution: Aqueous potassium sulfate and barium chloride undergo double displacement in aqueous solution to produce barium sulfate as a solid and leaving potassium chloride dissolved in the water.

$$K_2SO_4(aq) + BaCl_2(aq) \rightarrow BaSO_4(s) + 2KCl(aq)$$

This equation relates moles of reactants and products, which can be obtained easily from molarities.

$$\text{Moles } K_2SO_4 = 0.0346 \text{ L} \times 0.3004 \text{ moles/L}$$
$$= 10.40 \times 10^{-3} \text{ moles } K_2SO_4$$
$$\text{Moles } BaCl_2 = 0.0523 \text{ L} \times 0.0954 \text{ moles/L}$$
$$= 4.99 \times 10^{-3} \text{ moles } BaCl_2$$

Because 1 mole of K_2SO_4 is required per mole of $BaCl_2$, $BaCl_2$ is the limiting reagent. It follows that 4.99×10^{-3} moles $BaSO_4$ are produced. Converting to grams,

$$\text{Mass } BaSO_4 = 4.99 \times 10^{-3} \text{ moles} \times 233.386 \text{ g/mole}$$
$$= 1.165 \text{ g } BaSO_4.$$

Example: 31.05 mL of a 0.151 M solution of the acid, H_2SeO_4, is mixed with 58.74 mL of a 0.201 M solution of the base, NaOH. The acid and base react via proton transfer. Will the final solution be acidic (i.e., have an excess of H^+) or basic (i.e., have an excess of OH^-)?

Solution: The proton transfer reaction involving H_2SeO_4 and NaOH is given in reaction.

$$H_2SeO_4 + 2NaOH \rightarrow Na_2SeO_4 + 2H_2O$$

As in any reaction stoichiometry calculation, we must determine the numbers of moles of reactants that are initially present.

$$\text{mmoles } H_2SeO_4 = 31.05 \text{ mL} \times 0.151 \text{ m mole/mL}$$
$$= 4.689 \text{ m moles } H_2SeO_4$$
$$\text{mmoles NaOH} = 58.74 \text{ mL} \times 0.201 \text{ m mole/mL}$$
$$= 11.81 \text{ m moles NaOH}$$

(Here we have recognized that because a millimole is 1/1000 of a mole, and a mL is 1/1000 of a liter, molarity can be expressed either as moles/L or mmoles/mL.). The equation indicates that 2 mmoles NaOH are required to react with each mmole H_2SeO_4. There is more than enough NaOH available.

We conclude that H_2SeO_4 is the limiting reactant, and there will be some NaOH left over when reaction is complete. The final solution will contain an excess of OH-; it will be basic. Although we have answered the question at this point, we can take things one step further by determining the amount of NaOH remaining after reaction.

This is Amount unreacted,

NaOH = total NaOH – 2 × moles H_2SeO_4

= 11.81 mmole – 2 × (4.689 mmole)

= 2.432 mmoles

This quantity of NaOH is present in a total solution volume of 31.05 + 58.74 = 89.79 mL. The molarity of NaOH remaining is thus 2.432 mmole/89.79 mL = 0.0271 M

STOICHIOMETRY OF INCOMPLETE REACTIONS

We have made an implicit and drastic assumption about all of the chemical reactions that we have worked with thus far in the text: that they proceed until the limiting reactant is completely used up.

When this happens, chemists say that the reaction has "proceeded to completion." In practice, many reactions do not proceed to completion.

In most cases this is because the reaction reaches a state of chemical equilibrium in which both reactants and products are present.

Example: 5.25 moles of nitrogen and 12.00 moles of hydrogen are mixed and allowed to react according to equation. Calculate the amount of ammonia produced, assuming that the reaction proceeds only 20% of the way to completion, based on the limiting reactant.

$$N_2(g) + 3H_2(g) \rightarrow 2NH_3(g)$$

Solution: We must first determine the limiting reactant, based on the given starting amounts and the reaction stoichiometry.

The extent of reaction does not enter in to this determination, which is done in the usual way. In this problem, determination of limiting reagent is simple because the amounts are given in moles.

The equation requires 3 moles of H_2 for each mole of N_2 used, but only 12.00/5.25 = 2.29 moles of H_2 are available per mole of N_2. We conclude that H_2 is limiting. At this point we bring in the extent of reaction. Based on hydrogen, reaction proceeds only 20%. This means that 20% of the hydrogen initially present will be converted to ammonia, leaving 80% unconverted.

Moles H_2 that react = 0.20 × 12.00 = 2.40 moles

Moles H_2 left unreacted = 12.00 – 2.40 = 9.60 moles

The equation dictates that for each 3 moles of H_2 reacted, 2 moles NH_3 must form:

moles NH_3 formed = 2.40 moles H_2 × (2 moles NH_3/3 moles H_2) = 1.60 moles NH_3.

Incomplete reactions can be nicely handled in terms of a formalism that we describe as follows.

Step: Under the chemical equation, write the initial amounts of reactants and products present:

	N_2	$+ 3H_2 \rightarrow$	$2NH_3$
Initial	5.25	12.00	0

Zero is written under ammonia, because there is no ammonia initially present.

Step: Under the initial amounts, write the amounts of substances that are used (reactants) and formed (products). These numbers should be negative for reactants (because their amounts decrease during reaction), positive for products (because their amounts increase during reaction), and should be in accord with the stoichiometry of the reaction.

	N_2	$+ 3H_2 \rightarrow$	$2NH_3$
Initial	5.25	12.00	0
Reacts/forms		–2.40	

The number, -2.40, under H_2 is 20% of the initial amount; this is the amount of hydrogen that is used up. The number is negative to indicate this. What numbers should be placed under N_2 and NH_3? This is determined from reaction stoichiometry:

moles N_2 used = moles H_2 used × (1 mole N_2/3 moles H_2)
= –2.40 × (1/3)
= –0.80 moles N_2 moles NH_3 formed
= moles H_2 used × (2 moles NH_3/3 moles H_2)
= 2.40 × (2/3) = 1.60 moles NH_3

Enter these numbers in the "reacts/forms" line:

	N_2	$+ 3H_2 \rightarrow$	$2NH_3$
initial	5.25	12.00	0
reacts/forms	–0.80	–2.40	1.6

Step: Calculate the final amounts of the three substances by adding the amount reacted or formed to the initial amount:

	N_2	$+ 3H_2 \rightarrow$	$2NH_3$
initial	5.25	12.00	0
reacts/forms	–0.80	–2.40	1.6
final	4.45	9.60	1.60

This gives us a straightforward system for dealing with incomplete reactions.

OXIDATION STATES FROM LEWIS STRUCTURES

In an earlier section we presented some simple rules for the determination of oxidation state. In this section we show that it is possible to determine oxidation states of the atoms in covalent species (molecules and ions) directly from the Lewis structure of the species, using a procedure very similar to that used to obtain formal charge. Again, the procedure is systematic. It is carried through using the thiosulfate anion as an example.

Step: Develop an acceptable Lewis structure for the species, $S_2O_3^{2-}$. For $S_2O_3^{2-}$.

Step: Assign electrons to atoms according to these rules:

- Each atom is assigned all electrons of its lone pairs
- The bond electrons are assigned to the more

electronegative atom of the bond. If both atoms are of the same element, the bond electrons are divided in half.

The three oxygen atoms of $S_2O_3^{2-}$ are bonded similarly to the central sulfur atom. Each is assigned 6 electrons from lone pairs, and both electrons of its bond to S. Each oxygen is assigned 8 electrons. The terminal sulfur atom is assigned the 6 electrons of its lone pairs, and one half of the bond electrons for a total of 7. The central sulfur has no lone pairs, and is assigned only one bonding electron from the bond with the terminal sulfur.

Step: Compare the number of electrons assigned with the number of valence electrons that the atom normally has. The oxidation state is given as the difference between the normal number and the number assigned in the Lewis structure.

Subtracting: From the normal number, 6, gives 2– for the oxidation state of each oxygen atom.

Subtracting: From the normal number, 6, gives the oxidation state of the terminal sulfur atom as 1^-.

Subtracting: From the normal number, 6, gives the central sulfur an oxidation state of 6 1 = 5+.

Note that these differ from formal charges, which would be 1- for each oxygen; 1- for the terminal sulfur; and 2+ for the central sulfur. Applying the rules in section to the thiosulfate anion gives oxidation states 2- for oxygen and 2+ for sulfur. The sulfur value is the average of the values obtained from the Lewis structure.

The advantage of the Lewis structure method is that it produces individual atom oxidation states, rather than averages. You might try assigning oxidation states to the carbon atoms of acetic acid, C_2H_4O, using the Lewis structure approach.

You should find that the central carbon has an oxidation state of 3+, and the other carbon has oxidation state 3-. These average to zero.

Method for Determination of Stoichiometry

Job's Method, also called the Method of Continuous

Variation, is a simple and effective approach to the determination of chemical reaction stoichiometry. We will discuss it in the context of generic reaction,

$$aA + bB \Leftrightarrow dD$$

which can be rewritten in the form of by dividing all coefficients by "a".

$$A + kB \Leftrightarrow mD$$

where $k = b/a$ and $m = \Delta/a$. Job's method is based on the following fact: if a series of solutions is prepared, each containing the same total number of moles of A and B, but a different ratio, R, of moles B to moles A, the maximum amount of product, Δ, is obtained in the solution in which R = k (the stoichiometric ratio).

To implement Job's Method experimentally, one prepares a series of solutions containing a fixed total number of moles of A and B, but in which the R is systematically varied from large to small, and measures the amount of product obtained in each solution. One then plots amount of product versus R, and obtains a maximum at the initially-unknown value of k.

That the maximum amount of product should occur at the stoichiometric ratio can be justified both intuitively and mathematically. The intuitive justification runs as follows. When R is greater than k, there is an excess of reagent B, so reagent A is the limiting reagent. As R is systematically decreased toward k (i.e., as moles A increases and moles B decreases such that moles A + moles B stays constant) the amount of product increases with the amount of limiting reagent, A, until R becomes equal to k.

In contrast, when R is less than k, there is an excess of reagent A, and B is limiting. As R is systematically increased toward k (i.e., as moles B increases and moles A decreases), the amount of product increases with the amount of limiting reagent B, until R becomes = k.

Putting this all together, we see that as R is varied over the range from zero to the maximum value investigated, the amount of product obtained increases until R = k, then decreases as R becomes larger than k. This demonstrates Job's Method intuitively.

The mathematical justification is also quite simple. We use variable "x" to represent the moles of A in a particular solution, and assume that the total moles of A and B is to be kept at 1.0 throughout the series of solutions.

Then in each solution it will be true that:

$$x = \text{moles A}$$

$$1 - x = \text{moles B}$$

Our goal is to show that the maximum amount of product is obtained when R = moles B/moles A = $(1 - x)/x$ is equal to k. We approach this by finding the value of x that maximizes product.

According to equation, if x is less than the stoichiometrically correct amount of A, then A is limiting and moles product = mx. A plot of moles product versus x over a series of solutions should be linear, with slope m. Similarly, if x exceeds the stoichiometrically correct amount of A, then B is limiting and moles product = m$(1 - x)/k$.

A plot of moles product versus x over a series of solutions should also be linear, with slope = –m/k. The first plot will proceed up to the right as x increases.

The second plot will proceed down to the right as x increases. At some point then, the two straight lines will intersect. At the intersection, they have a point in common. The value of x corresponding to this point is obtained by equating the ordinate values and solving for x:

$$mx = m(1 - x)/k$$

Solving gives $x = 1/(1 + k)$.

Substituting in the expression for R, $(1 - x)/x$, we find that the two lines intersect when

$$R = (1 - x)/x = \{1 - [1/(1 + k)]\}/[1/(1 + k)] = k.$$

Because the amount of product increases as k is approached from either direction, the point of intersection of the lines occurs at the maximum amount of product obtainable. We have therefore shown that maximum product is obtained when $R = k$. This is what we set out to demonstrate.

Example: A and B are known to react to form Δ, but the stoichiometry is uncertain. A Job's Method study yields the following data. Plot quantity of product versus moles A to determine the stoichiometry.

moles A	moles B	grams product
0.2	1.8	2.5
0.3	1.7	3.75
0.4	1.6	5.0
0.6	1.4	4.38
0.8	1.2	3.75
1.0	1.0	3.12

Solution: The value of k is clearly 4, because at the maximum, moles B/moles A = 1.6/0.4 = 4.

A Detailed Example: Stoichiometry of Copper Amino Acid Complexes by Job's Method

We now describe the application of Job's method to the determination of the stoichiometry of amino acid complexes with the Cu^{2+} ion. Job's Method may be applied to stoichiometric detemination in either of 2 ways. For clarity, we discuss it in terms of the general reaction:

$$: A + xB \rightarrow C$$

Table. The Standard Approach

Solution Number	Volume 0.1 M A solution, mL	Volume 0.1 M B solution, mL	Volume additional solvent, mL	Ratio [B]/[A]	mmoles Product
1	0	0.9	1.1	infinite	0
2	0.1	0.8	1.1	8	0.01
3	0.2	0.7	1.1	3.5	0.0175
4	0.3	0.6	1.1	2	0.015
5	0.4	0.5	1.1	1.25	0.0125
6	0.5	0.4	1.1	0.8	0.01
7	0.6	0.3	1.1	0.5	0.0075
8	0.7	0.2	1.1	0.28	0.005
9	0.8	0.1	1.1	0.125	0.0025
10	0.9	0	1.1	0	0

The Standard Method. A series of solutions is prepared in which the sum of the number of moles of A and the number of moles of B is kept constant, but in which the relative amounts of A and B are systematically varied.

Thus moles B/moles A varied from 0 to some value certain to be larger than x. For example, the following series of solutions might be prepared from two stock solutions, one 0.1 M in A and the other 0.1 M in B: Once the solutions are prepared, the amount of product formed is measured in each solution, and a plot of the amount of product formed versus moles of B in the solution is constructed.

The crux of Job's Method is that the maximum amount of product will form when A and B are present in the correct stoichiometric ratio.

Normally the plot looks like an inverted "V", with the vertex skewed toward one side or the other. Straight lines are drawn through the points on both sides of the vertex, and the moles of B corresponding to their intersection point allows the stoichiometry to be determined. The value of x is then calculated from equation:

x = (moles of B at intersection)/[total moles
$A + B$ moles of B at intersection]

Because the intersection in the plot occurs at moles B = 0.072/(0.09 – 0.072) = 4, the coefficient of reactant B in equation is 4. The challenging aspect of the application of Job's Method is in finding a way to measure the amount of product formed in each solution.

If the product precipitates, the product in each solution may be recovered by filtration and weighed. If the product absorbs light at a known wavelength, then each solution can be placed in a spectrometer and the absorbance at that wavelength determined.

There are many other possibilities. The primary restraint, particularly when using a spectrometric method, is that only the amount of product is measured.

If, for example, one of the reactants also has significant absorbance at the wavelength being used to monitor the amount of product, the shape of the Job's method plot will be affected, sometimes to an extent that prevents determination of the stoichiometry in a straightforward way. It is important to be alert for this problem.

The Limiting Reagent Method. In this modification of Job's Method, a series of solutions is prepared containing a fixed number of moles of reactant A and varying amounts of reactant B, such that the ratio, moles B/moles A runs the gamut from 0 to a value known to be larger than *x*.

The amount of product in each solution is then measured. Once the amount of B exceeds the stoichiometrically required amount, A is the limiting reagent, and the amount of product formed will remain constant. Thus by noting the ratio [B]/[A] at which the amount of product formed levels off, the stoichiometry can be determined. A plot of amount of product versus [B]/[A] will reveal this point visually. Table provides data for a series of solutions.

Table. The Limiting Reagent Approach

Solution Number	Volume 0.1 M A solution, mL	Volume 0.1 M B solution, mL	Volume additional solvent, mL	Ratio [B]/[A]	mmoles Product
1	0.15	0	1.85	0	0
2	0.15	0.05	1.8	0.33	0.00125
3	0.15	0.1	1.75	.667	0.0025
4	0.15	0.15	1.7	1	0.00375
5	0.15	0.2	1.65	1.33	0.005
6	0.15	0.25	1.6	1.67	0.00625
7	0.15	0.3	1.55	2	0.0075
8	0.15	0.45	1.4	3	0.01125
9	0.15	0.60	1.25	4	0.015
10	0.15	0.75	1.1	5	0.015
11	0.15	0.90	0.95	6	0.015

Chapter 9

Principles of Chemical Dynamics

OVERVIEW

Some Basic Ideas. Until now, our major concern has been whether or not a chemical reaction of interest may or may not occur under specified conditions of temperature, pressure, and concentration. The answer to this question is in the realm of thermodynamics. If thermodynamics tells us that a reaction may occur, then we must be concerned with how rapidly and by what pathway it will occur, and how we may influence this.

The speed (rate) at which a reaction takes place is a matter of chemical dynamics, or alternately, chemical kinetics. Chemical dynamics (kinetics) is the study of the rates (speeds) and mechanisms (pathways) of chemical reactions.

We have briefly encountered the rate concept before, when we discussed the manner in which vapour pressure changes with time, and when we presented dynamic equilibrium as a situation of balanced, opposing rates. Although chemical kinetics is a well-developed field experimentally, we are still a long way from the theoretical grasp of rate that we have of equilibrium.

Thus our predictive ability in the realm of kinetics is very limited. Thermodynamics may predict that a reaction should proceed essentially completely. Whether it does so depends on the rate with which it occurs. For example, consider the reaction below at 298 K:

$$C(graph) + O_2(g) \rightarrow CO_2(g) \ [DG^o_R = -394.4 \text{ kJ}]$$

Thermodynamics predicts that the reaction should go to completion. But carbon (coal) deposits exist in the atmosphere without noticeably reacting; and the "lead" in our pencils does not burst spontaneously into flame. The reaction is favored, but is very slow at 298 K.

We say that C(graph) is thermodynamically unstable in the presence of $O_2(g)$ at 298 K, but is kinetically inert. As a second case, consider the acid-base neutralization reaction:

$$H_3O^+(aq) + OH^-(aq) \rightarrow 2H_2O(l)\ [DG^o_R = -58.28\ kJ]$$

Thermodynamics predicts that this reaction, too, should go to completion. In fact, it occurs almost instantaneously; equilibrium is attained as rapidly as the reactants are mixed. Thus we say that H_3O^+ is both thermodynamically unstable and kinetically labile in the presence of OH^-. Two sets of descriptive words are necessary to distinguish thermodynamics and kinetics:

Thermodynamics: stable $DG > 0$; unstable $DG < 0$

Kinetics: inert (reaction is slow); labile (reaction is fast).

As we have seen above, there is no necessary correlation between "stable" and "inert", or "unstable" and "labile". This lack of correlation reflects the fact that thermodynamics deals with values of state functions in various equilibrium states of a system; kinetics deals with the rate at which a path between states is traversed. Reaction rates are extremely important in many ways in our lives. Sustained muscle action depends on the continuous supply of energy produced by glucose combustion in the mitochondria of cells. Energy production must be fairly rapid, otherwise sustained action is not possible.

The extremely rapid (explosive) burning of gasoline in the internal combustion engine provides the motive power for automobiles. Food cooks (or, as the chemist would say, oxidizes) extremely slowly at room temperature, but in a relatively short time at more elevated temperature. Our first concern must be to discuss the manner in which reaction rates are expressed, and to introduce some of the terminology of chemical dynamics.

What is a rate? In general, a rate is a number expressing how much a quantity changes in a convenient interval of time. Thus an interest rate is the amount by which a sum of money increases in 1 year; travel rate in an automobile or airplane is the distance covered in 1 hour; heart rate is the number of heartbeats per minute; thus a rate is the change in a quantity per unit time:

$$\text{Rate} = \text{Dquantity/Dt}$$

Note that the units of rate must be "quantity/time, or "quantity-time^{-1}". Thus interest rate has units of dollars/year; travel rate, miles/hour; and heart rate; beats/minute.

How are we to express the rate of a chemical reaction? To develop the answer to this question, we will begin with the simple chemical process in equation. The equation describes quantitatively the conversion of ozone to oxygen that occurs in the earth's upper atmosphere.

$$2O_3(g) \rightarrow 3O_2(g)$$

When this (or any other) reaction takes place, it is the amounts of reactants and products that change. As we know, amount can be expressed in either mass, moles, or molecules. The coefficients in the chemical equation relate both molecules and moles of reactants and products but do not directly relate masses. So we will agree that our measure of amount should be either molecules or moles. Further, since work in the laboratory is carried out on the macroscopic scale, moles is the more suitable unit of amount.

So we might describe the rate of process either in terms of the number of moles of O_3 used or of O_2 produced per unit time. However, there is a major difficulty with this approach. The number of moles of ozone used will depend upon the volume of atmosphere that we choose to examine!

We cannot specify a rate without also specifying the volume examined. Rather than use moles/time, then, we choose to divide by the appropriate volume to give moles/liter-time, or concentration/time. Rates expressed in these units are independent of the actual amount of reaction carried out. At this point then we can express the rate of reaction in either of two ways:

$$\text{Rate} = \text{Moles } O_3 \text{ used per liter per time} = -\Delta(\text{conc } O_3)/\text{Dtime}$$

Rate = Moles O_2 produced per liter per time = Δ(conc O_2)/Dtime

The units of the rate are then moles per liter per time, where the time unit should be the one most convenient for the reaction being measured. Chemists agree that the rate of a reaction should be expressed as a positive number.

As reaction proceeds, the concentration of ozone becomes smaller while that of oxygen becomes larger. Thus the change in ozone concentration, D(conc O_3), is a negative quantity. To produce a positive rate, it is necessary to place the minus sign preceding the D(conc) term, as has been done. In general, whenever the rate of a reaction is expressed in terms of the change in concentration of a reactant, D(conc reactant) is negative and the minus sign is necessary. The minus sign is unnecessary because D(conc O_2) is positive.

There is an additional difficulty, though. We would like to specify a single number representing the rate of a reaction under given conditions.

However, the two expressions above give different numbers for the rate, because dioxygen appears 1.5 times faster than ozone disappears (i.e., in the same time that 2 moles of ozone react, 3 moles of dioxygen appear).

This can be fixed by normalizing the two expressions in with the stoichiometric coefficients. Dividing D(conc O_3) by the coefficient 2 gives the same number as dividing D(conc O_2) by the coefficient 3:

Rate = –1/2 × D(conc O_3)/Dtime = 1/3 × D(conc O_2)/Dtime

Replacing "conc O_3" with the brackets to represent molar concentratration:

$$\text{Rate} = -1/2 \times \Delta[O_3]/\Delta t = 1/3 \times \Delta[O_2]/\Delta t$$

The symbol, D, signifies a finite (i.e., measurable) change in a quantity.

In the limit as the length of the time interval, Dt, approaches zero, we replace D(conc) and Dt with the differential (infinitesimal) quantities D(conc) and dt to obtain the differential expression of rate.

$$\text{Rate} = -1/2\ \Delta[O_3]/dt = 1/3\ \Delta[O_2]/dt$$

The appropriate formulation for finite difference methods of analysis.

For the completely general chemical reaction, proceeding from left to right from reactants, *A* and *B*, to products, D and *F*, the rate of reaction can be represented using any one of the expressions:

$$aA + bB \rightarrow d\Delta + fF$$

$$\text{Rate} = -1/a\ \Delta[A]/dt = -1/b\ \Delta[B]/dt = 1/\Delta\ \Delta[\Delta]/dt = 1/f\ D[F]/dt$$

As discussed earlier, negative signs are necessary when rate is expressed in terms of disappearance of a reactant; and division by the stoichiometric coefficient guarantees that the same numerical value of rate is obtained from all four expressions.

Example: In the reaction of ozone to produce oxygen, it is found that under certain conditions of temperature and concentration, 0.0360 moles of ozone per liter react in a 2-hour period. How much dioxygen is produced in this time period? What is the average rate of reaction over this time period?

Solution: The chemical equation, tells us that 3 moles of O_2 are produced for each 2 moles of ozone reacted.

Therefore,

0.0360 moles ozone × (3 moles O_2/2 moles O_3) = 0.0540 moles dioxygen are produced.

The average rate of reaction can be expressed in either of two ways:

$$\text{Rate} = -1/2\ \Delta[O_3]/\Delta t = -1/2\ (-0.0360\ \text{moles/L})/2\ \text{hours} = 0.009\ \text{moles/L-hour}$$

$$\text{Rate} = -1/3\ \Delta[O_2]/\Delta t = 1/3\ (0.0540\ \text{moles/L})/2\ \text{hours} = 0.009\ \text{moles/L-hour}$$

Both expressions give the same positive rate of reaction.

Example: The rate was studied at 55°C by measuring the concentration of t-butyl bromide (*t*-BuBr) as a function of time. The data acquired are given below. Use the data to estimate the rate of reaction 20 seconds after reaction was begun.

$$t\text{-BuBr} + H_2O \rightarrow t\text{-BuOH} + HBr$$

Time, s	[*t*-BuBr]
0	0.100
10	0.0876
20	0.0768
30	0.0672
40	0.0590
50	0.0517
60	0.0453
80	0.0348
100	0.0267
120	0.0205
180	0.0093
240	0.0042

We begin by plotting the data to obtain a picture of the manner in which the concentration of t-BuBr varies with time. A plot of [*t*-BuBr] versus time is given.

The shape of the plot is typical of a rate process: the concentration of reactant changes very rapidly at first, but as time goes on, the change in concentration per unit time becomes less and less as the concentration of *t*-BuBr approaches its final (equilibrium) value. We have seen this characteristic time plot before in our discussions of vapour pressure and approach to chemical equilibrium. Most concentration-time plots for a reaction in progress have this same general appearance.

Two approaches are possible for obtaining the rate at the 20-second mark of the reaction. The finite difference expression of rate can be used to calculate the average rate during the time interval preceding or following the 20-second mark.

Both average rates are calculated below:

Average rate between 10 and 20 seconds:

$= -(\text{conc at 20 s} - \text{conc at 10 s})/(20 - 10)\ Ms^{-1}$

$= -(0.0768 - 0.0876)/(20 - 10)$

$= 1.08 \times 10^{-3}$ moles $L^{-1}\ s^{-1}$

Average rate between 20 and 30 seconds:

$= -(0.0672 - 0.0768)/(30 - 20)$

$= 0.96 \times 10^{-3}$ moles $L^{-1}\ s^{-1}$

The average rate over the 20 – 30 second time interval is less than that over the preceding interval because the reaction slows down as reactants are used up.

This slowing down is readily apparent in the shape of the plot. However, the values are not too much different, so either may be used as an estimate of the rate of reaction at the 20 second mark.

The second approach is to use the differential expression of rate. Just as *dy/dx* at $x = x_o$ is the slope of a plot of *y* versus *x* at $x = x_o$, Δ[*t*-BuBr]/*dt* at a particular time, t_o, is the slope of a plot of [*t*-BuBr] versus time at the particular time, t_o. This slope is the same as the slope of the tangent to the curve at t_o. The tangent at t = 20 s is shown in the figure. It's slope is readily calculated to be 1.01×10^{-3} moles L^{-1} s^{-1}.

As we might have expected, this value gives a rate intermediate between the values obtained above, based on finite time intervals.

Because the slope of the tangent at a particular time represents the rate of reaction at that time, the rate obtained in this manner is called the instantaneous rate at the particular time.

It is analogous to the speed that you measure when you glance at your automobile speedometer at a particular instant. As we have shown in this example, the instantaneous rate can be obtained by constructing tangents; or it can be estimated using the average rate over either the preceding or succeeding interval of time.

We can summarize our introductory discussion of reaction rate as follows:

- From a plot of [reactant] or [product] versus time, we can measure the rate at any particular time during the reaction as the slope of the tangent to the curve at that time; the initial rate is the slope of the tangent at t = 0.
- Rate decreases as the reaction proceeds toward equilibrium; this is evident from the shape of the concentration-time plot.
- The units of rate are []/time, or moles/L-time.

THE RATE LAW

The rate of a chemical reaction depends on a number of factors, some of which are listed below:

- The molecular structure(s) of the reactant(s)
- The medium in which the reaction is carried out (for example, does the reaction occur in the gas phase, or does it occur with reactants dissolved in a solvent?)
- Temperature
- The presence or absence of a catalyst
- The concentrations of the reacting species

The dependence of rate on molecular structure is the most obvious but least understood factor. This complex matter is a focus of intense laboratory research.

At present, it is not possible to summarize this dependence in a neat, clean package. In fact, this incompletely understood factor is what prevents predictive ability in kinetics. For molecules to react, they must directly interact; to directly interact, they must collide.

This leads us to expect that rate should be proportional to the frequency of molecular collisions. In turn, the frequency of collisions should depend on the number of molecules per unit volume, which is the concentration. This concentration dependence of rate is indeed observed experimentally.

For most chemical reactions, the dependence of reaction rate on the concentrations of the reacting species is surprisingly simple. In almost all cases, the reaction rate is found to be proportional to the product of the concentrations of the reactants, each concentration raised to an integer power that is almost always 1 or 2.

Thus for the general reaction, the rate law is expected to take the form:

$$\text{Rate} = -1/a\ \Delta[A]/dt = k\ [A]^x\ [B]^y$$

Equation is called the differential rate law, because it involves the ratio of differentials (derivative), D[*A*]/*dt*. The exponents, *x* and *y*, are integers, and for most reactions are either 1 or 2. It must be stressed that IN MOST CASES *x* and *y* are different from the stoichiometric coefficients, *a* and *b*; it is not valid to write the rate law for a reaction by raising the concentrations to the stoichiometric powers, unless the reaction is an elementary process.

We will have more to say about elementary processes in a subsequent section. The integer, x, is called the order of the reaction in reactant A; y is the order in B. The sum, $x + y$, is called the overall reaction order.

The reaction orders must be determined by experiment; they can not be written down simply by looking at the reaction stoichiometry.

To illustrate this statement for a specific case, the reaction below, involving 31 moles of reactants, has the accompanying rate law. This very complex overall reaction has a simple rate law, which is first order in each reactant, and second order overall.

$$30C_2H_5OH + B_{10}H_{14} \rightarrow 10B(OC_2H_5)_3 + 22H_2(g)$$

$$\text{Rate} = k[C_2H_5OH][B_{10}H_{14}]$$

Finally, k (and in the expression just above) is the proportionality constant relating Rate to the concentration product.

It is called the rate constant for the reaction. The rate constant contains implicit within it information about the dependence of rate on molecular structure, reaction medium, and temperature.

A word about the units of the rate constant is in order. Generally, k has whatever units are required to give the rate units of concentration/time.

Example: Shows how units for the rate constant may be determined for various rate laws.

Example: For each rate law below, state the order of the reaction in each reactant, and determine the units of the rate constant, k.

$-\Delta[A]/dt = k$

$-\Delta[A]/dt = k[A]$

$-\Delta[A]/dt = k[A]^2$

$-\Delta[A]/dt = k[A][B]$

$-\Delta[A]/dt = k[A]^2[B]$

Solution:

Rate Law	Reaction Order	Units of k
$-\Delta[A]/dt = k$	Zero order (no concentration dependence)	k has units of rate, Moles L^{-1} $time^{-1}$
$-\Delta[A]/dt = k[A]$	First order in A	Moles/L-time = k(moles/L) so k has units of $time^{-1}$
$-\Delta[A]/dt = k[A]^2$	Second order in A	Moles/L-time = k(moles/L)^2 so k has units L $moles^{-1}$ $time^{-1}$
$-\Delta[A]/dt = k[A][B]$	First order in B	First order in A Same as for the previous case Second order overall
$-\Delta[A]/dt = k[A]^2[B]$	Second order in A First order in B third order overall	Moles/L-time = k(Moles/L)^3 so k has units $L^2 moles^{-2}$ $time^{-1}$

Some Specific Examples of Rate Laws. Zero-order Reactions. Chemical reactions having overall order zero are rare. However, evaporation of a pure liquid, is an example of a physical process with a zero-order rate law. Recall that the rate of evaporation of a liquid having exposed surface area, A, is just:

$$\text{Rate} = kA$$

As long as the exposed surface area remains constant, we may rewrite this as:

$$\text{Rate} = k', \text{ where } k' = kA$$

Since this rate is independent of concentration of liquid, the process is zero order in liquid.

First-Order Reactions

Reactions and physical processes exhibiting first order rate laws are very common. Condensation is a first order physical process. The rate of condensation of vapour to liquid in a closed system is given by:

$$\text{Rate} = k\, P_{vap}$$

where P_{vap} is the vapour pressure of the substance. However, P is related to concentration in moles/L by the ideal gas law, leading to:

$$\text{Rate} = -\Delta[\text{vapour}]/dt = k \times [\text{vapour}]$$

Molecular rearrangements (often called isomerization reactions) frequently follow first order rate laws. Vision is critically dependent upon cis-retinal → trans-retinal where cis- and trans- retinal contain the same atoms, bonded together in the same way, but with somewhat different spatial arrangements. The rate of conversion of cis- to transform is given.

$$-\Delta[\text{cis}]/dt = k[\text{cis}]$$

Similarly, trans- converts to cis- by a first order rate law with a different rate constant. Finally, nuclear decay processes are first order. Thus ^{238}U decays according to equation, which is governed by the rate law.

$$^{238}U \rightarrow {}^{234}Th + {}^{4}He$$

$$\text{Rate} = -\Delta(\text{mass } ^{238}U)/dt = k\,(\text{mass } ^{238}U)$$

The rate constant for this process has the extremely small value, 1.54×10^{-10} year^{-1}.

The rate constant, k, for a first-order process has units of time^{-1}. It's reciprocal, $1/k$, thus has units of time. The quantity, $1/k$, is given the symbol t, and is called the time constant for the reaction. This is a measure of the time required for the reaction to take place.

The time constant for decay of ^{238}U is $1/k = 6.49 \times 10^9$ years. The fact that this time constant is so large is the reason why ^{238}U is still found in the earth's crust, which is estimated to be about 4.5 billion years old.

Example: The time constant for the radioactive isotope, ^{226}Ra, is 2338 years. Calculate the number of atoms of ^{226}Ra that decay each minute in a 2.00 g sample of ^{226}Ra.

Solution: Nuclear decay is a first order process. We can therefore write the following rate law for decay of ^{226}Ra:

Rate $= -D(^{226}Ra \text{ atoms})/dt = k(\text{Number of } ^{226}Ra \text{ atoms})$

$-dN/dt = kN$ where N is the number of ^{226}Ra atoms in the sample, and

$k = 1/t = 4.28 \times 10^{-4}$ year^{-1}.

$N = (2.00g/226.025 \text{ g/mole}) \times 6.023 \times 10^{23}$ atoms/mole

$= 5.33 \times 10^{21}$ atoms.

$-dN/dt = (4.28 \times 10^{-4} \text{ year}^{-1})(5.33 \times 10^{21} \text{ atoms})$

$= 2.28 \times 10^{18}$ atoms/year

There are 5.256×10^5 minutes in 1 year.

Therefore,

$-dN/dt = (2.28 \times 10^{18} \text{atoms/year})/(5.256 \times 10^5 \text{ minutes/year})$

$= 4.34 \times 10^{12}$ atoms/minute.

Second Order Processes

Many reactions exhibit second-order rate laws, either simple (second-order in a single reactant) or mixed (first-order in each of two reactants). Here we consider two specific reaction types that exhibit second-order kinetics. Dimerization reactions, in which two molecules unite to form a larger molecule, obey simple second-order kinetics.

The dimerization of NO_2 is shown below:

$$2NO_2\,(g) \rightarrow N_2O_4\,(g)$$

The rate law:

$$-\Delta[NO_2]/dt = k[NO_2]^2$$

where the second order rate constant is — $M^{-1}s^{-1}$ at °C. Other examples of dimerization processes are the formation of a Cl_2 molecule from two Cl atoms and the formation of one molecule of C_4H_8 from two molecules of C_2H_4.

In general, a reaction of the general type is expected to obey simple second-order kinetics:

$$2A \rightarrow A_2$$

Adduct formation processes, such as Lewis acid-base reactions and the formation of complexes involving enzymes and their substrates, are also second-order processes. Most such processes occur very rapidly. The role of enzyme-substrate complexes in enzyme catalysis will be discussed in a subsequent section.

"Net" Reaction Rate

The experimental approaches to the determination of rate laws for reactions. To preface that discussion, we briefly discuss here the concept of net reaction rate.

Except in some cases, we can experimentally measure only the net rate of a reaction, where the net rate is defined as:

Net reaction rate = rate of forward reaction – rate of reverse reaction,

$$\text{Rate}_{\text{net}} = \text{Rate}_{\text{forward}} - \text{Rate}_{\text{reverse}}$$

The general reaction is reproduced below, this time acknowledging that in general the reverse process also occurs:

$$aA + bB \Leftrightarrow dD + fF$$

For this reaction we expect the forward rate to depend only on the concentrations of reactants, *A* and *B*, and the reverse rate to depend only on the concentrations of products, D and *F*:

$\text{Rate}_{\text{forward}} = k_f [\text{A}]^x [B]^y$

$\text{Rate}_{\text{reverse}} = k_r [\text{D}]^p [F]^q$

The net rate is therefore expected to be a potentially complex function of the concentrations of reactants AND products. The experimentally determined rate constant is a composite function of the forward and reverse rate constants, k_f and k_r.

$\text{Rate}_{\text{net}} = k_f [A]^x [B]^y - k_r [D]^p [F]^q$

Once reaction reaches equilibrium, the net rate will of course be zero. Applying this idea to, our familiar criterion for dynamic equilibrium:

$\text{Rate}_{\text{net}} = 0 = \text{Rate}_f - \text{Rate}_r$

$\text{Rate}_f = \text{Rate}_r$

There are a few situations in which the net rate is equal to the forward reaction rate:

- One of the products escapes from the reaction system, so that its concentration remains at or very near zero at all times. A gaseous product thai escapes from a solution is an example of this situation.

- The reaction has a large value of K_{eq}, so that the reverse rate is negligible except near equilibrium, which is not achieved until almost all of the limiting reactant has been used.
- At the very beginning of a reaction in which only reactants are initially present. At time zero, the reverse rate must be zero because no products are present yet.

RATE LAW THE INITIAL RATE METHOD

When applicable, the method of initial rates is a simple and clean way to obtain the rate law. The initial rate of a reaction is the instantaneous rate at $t = 0$, when reactants are first mixed together and begin to react. In a plot of reactant concentration versus time, it is given by the slope of the tangent at $t = 0$.

The slope of this type of plot is always steepest at the beginning; the initial rate is thus the fastest rate that we observe during the time course of the reaction process. One major advantage of the initial rate method is that the reverse rate is zero, provided only reactants and NO products are initially present.

Thus only the forward rate contributes to the initial rate, and complications resulting from the reverse reaction are avoided. The initial rate law depends only on concentrations of reactants. The method of initial rates is best illustrated by example.

Example: The initial rate was measured at several combinations of initial reactant concentration for the following reaction:

$$A + 2B + 3C \rightarrow \Delta + E$$

The rate law is expected to have the following general form:

$$-\Delta[A]/dt = k[A]^x[B]^y[C]^z$$

The following data were obtained. Use the data to determine the values of x, y and z.

Run Number	$[A]_o$, M	$[B]_o$, M	$[C]_o$, M	Initial Rate, M/s
1	0.05	0.10	0.10	6.00×10^{-3}
2	0.05	0.15	0.10	13.5×10^{-3}
3	0.05	0.15	0.05	13.5×10^{-3}
4	0.10	0.10	0.05	12.0×10^{-3}

Solution: The key to the initial rate method is to examine the dependence of rate on one reactant at a time, using data obtained when the other reactants do not vary in initial concentration. In the first two data entries above, the initial concentrations of both *A* and C are the same in both runs; only the initial concentration of *B* is changed.

These two data items can therefore be used to determine the order of the reaction in *B*. We see that when the initial concentration of *B* is increased from 0.10 to 0.15, an increase of a factor of 1.5, the rate increases from 6 to 13.5×10^{-3} moles/L–s, which is a factor of 2.25. We translate this information into equation form:

$$\text{Rate 2/Rate 1} = [B]_2{}^y/[B]_1{}^y = ([B]_2/[B]_1)^y$$

$$2.25 = (1.5)^y$$

We recognize that *y* must be 2. The reaction is therefore second order in *B*, and the rate law thus far is

$$-\Delta[A]/dt = k[A]^x[C]^z[B]^2$$

In runs 2 and 3, the initial concentrations of *A* and *B* are the same, while that of C is changed. We see from these data that even though the concentration of C is only half as large in run 3 as in run 2, the rate is unaffected. We conclude that the rate does not depend on [C], hence that $z = 0$.

The rate law to this point is:

$$-\Delta[A]/dt = k[A]^x[B]^2$$

There are no two kinetics runs in which only [*A*] is varied. However, runs 1 and 4 can be used to obtain the order in *A* because we have just determined that rate does not depend on [C]. Runs 1 and 4 indicate that the rate doubles when the initial concentration of *A* doubles.

The order in A is thus 1, and the final rate law is:

$$-\Delta[A]/dt = k[A][B]^2$$

Once the rate law is known, the rate constant may be calculated from any one of the individual runs. Choosing the first run,

$$k = \text{Rate}/[A][B]^2 = 6.00 \times 10^{-3}/(0.05)(0.10)^2 = 12.0\ \text{L}^2/\text{mole}^2\text{s}$$

As this example shows, the initial rate method, when it is applicable, provides a clean, simple path to the rate law. The italicized qualifier is important, however; in practice, the method of initial rates is only rarely used, for several reasons. In most kinetics studies, the chemist examines the manner in which the concentration of a selected reagent varies with time. The resulting data are presented either in a table, or in a plot of concentration versus time.

The data obtained at the beginning of the reaction, where the rate is largest and the time interval is smallest, is usually less reliable than that obtained at later times, when the rate is less. It is undesirable to base an analysis on one's least reliable data. Second, and more important, the method of initial rates uses only a small fraction of the concentration time data; the rest is wasted.

Third, to apply the initial rate method as was done in the example above requires that full concentration-time profiles be collected for each run, in order to extract the slope of the concentration-time plot at $t = 0$. Collection of this much data requires a lot of experimental effort. Most of the resulting data is then ignored if the initial rate method is used. For these reasons and others, chemists have developed other methods for determining rate laws from data, which make use of all of the data collected. Some of these will be discussed now.

Integration of the Rate Law

It is possible to convert the differential rate law into an integrated form, in which concentration is expressed directly as a function of time. The concentration-time data obtained during a kinetics run in the lab may then be matched against the function. If the data conform to the function, it may be concluded that the reaction follows the corresponding differential rate law.

Integration is very easy for rate laws of the form:

$$-\Delta[A]/dt = k[A]^n$$

We will examine the first and second-order cases, with $n = 1$ and 2 respectively.

The first order rate law may be rearranged to put the concentration terms on one side and time on the other side of the equals sign:

$$-\Delta[A]/[A] = kdt$$

Both sides of the equation are then integrated over the limits, $[A] = [A]_o$ at $t = 0$ and $[A] = [A]_t$ at some later time t:

$$\int_{[A]0}^{[A]t} d[A]/[A] = k\int_0^t dt$$

The integral of dx/x is ln x, and the integral of dx is x. Evaluating over the limits gives:

$$\ln([A]_{t/}[A]_o) = -k\,t$$

This is the so-called integrated first order rate law. For a reaction following simple first order kinetics, a plot of ln $[A]_t$ versus t is linear, with slope $-k$. Thus if concentration-time data for a reaction are plotted according to equation, and linearity is obtained, it may be concluded that the reaction follows a simple first-order differential rate law. An alternative expression is obtained by exponentiating both sides:

$$[A]_t = [A]_o e^{-kt}$$

Here $e = 2.718...$ is the base of natural logarithms. Substituting the reciprocal of the time constant, t, for the rate constant gives:

$$[A]_t = [A]_o e^{-t/t}$$

Thus when an interval of time equal to the time constant has elapsed, the concentration of A will have decreased to $1/e$ (about 1/3) of its initial value.

With the integrated first-order rate law in hand, we are in a position to consider a quantity called the half-life, $t_{1/2}$, which is similar but not identical to the time constant. The half-life for a process is defined as the time required for the concentration of the limiting reactant to decrease to 1/2 of its initial value. An expression for the half life is easily obtained from by realizing that $[A]_t = [A]_o/2$ at $t = t_{1/2}$. Making these substitutions leads to the following expression for the half-life of a first-order reaction:

$$t_{1/2} = \ln 2/k = 0.693/k$$

For $n = 2$, corresponding to a second-order process, rearrangement leads:

$$-\Delta[A]/[A]^2 = k\,dt$$

Integration of both sides of this equation results:

$$1[A] - 1/[A]_o = k\,t$$

For a reaction with a simple second order rate law, a plot of concentration-time data as $1/[A]$ versus t is linear, with slope k. Thus a reaction for which the concentration-time data produce a linear $1/[A]$ versus t plot may be concluded to follow a second-order rate law.

Example: Plot the concentration-time data in Example. according to equations to determine whether the hydrolysis of *t*-butyl bromide obeys a simple first- or second-order rate law.

Solution: The concentration-time data from Example.are reproduced below. Columns have been added for ln [*t*-BuBr] and 1/[*t*-BuBr]

Time, s	[t-BuBr], moles/L	ln [t-BuBr]	1/[tBuBr]
0	0.100	–2.303	10
10	0.0876	–2.435	11.42
20	0.0768	–2.567	13.02
30	0.0672	–2.700	14.88
40	0.059	–2.830	16.95
50	0.0517	–2.962	19.34
60	0.0453	–3.094	22.08
80	0.0348	–3.358	28.74
100	0.0267	–3.623	37.45
120	0.0205	–3.887	48.78
180	0.0093	–4.678	107.53
240	0.0042	–5.473	238.10

The linearity of the first plot indicates that the reaction is first-order in *t*-BuBr. The rate constant, obtained from the slope is 1.32×10^{-2} s^{-1} (recall that for a first order reaction, the units of the rate constant must be $time^{-1}$, resulting from the division of rate in conc/time by concentration).

Example: One product resulting from fission of ^{235}U in nuclear reactors is ^{137}Cs, a radioactive nucleus with a time constant of 43.3 years. How much time will it take for the radioactivity level of this waste to fall to 1% of its initial value, which is considered a safe level?

Solution: This problem is readily solved using equation:

$$\ln [A]/[A]_o = -kt = -t/t$$

$$t = -\ln(0.01)t = -\ln(0.01)(43.3) = 200 \text{ years}$$

The waste must be safely stored for 200 years before it will be safe.

Finite Difference Methods of Numerical Analysis

Another approach to determination of the rate law and rate constant from concentration-time data is to calculate the average rate over each experimental time interval.

The average rates can then be plotted in turn against concentration, concentration squared, and so on until a linear plot is obtained.

This is a relatively simple way to determine whether a reaction is first-order or simple second order (that is, has a rate law of the form, Rate = $k[A]^2$).

Estimating derivatives (slopes) and integrals (areas) using finite changes in variables is called the Method of Finite Differences. We illustrate this method using the data for hydrolysis of *t*-BuBr from Examples.

Example: Use finite difference methods to obtain the rate law and rate constant for the hydrolysis of *t*-butyl bromide according to equation.

Solution: The time and concentration data from Example are reproduced below. In addition, values of *Dt*, D[*t*-BuBr], and rate = $\Delta[t\text{-BuBr}]/\Delta t$ have been calculated for each pair of successive data points.

Time, s	Dt, s	[t-BuBr], moles/L	D[t-BuBr], moles/L	Rate × 10^3
0		0.100		
10	10	0.0876	–.0124	1.24
20	10	0.0768	–.0108	1.08
30	10	0.0672	–.0096	0.96
40	10	0.0590	–.0082	0.82
50	10	0.0517	–.0073	0.73
60	10	0.0453	–.0064	0.64
80	20	0.0348	–.0105	0.525
100	20	0.0267	–.0081	0.40
120	20	0.0205	–.0062	0.31
180	60	0.0093	–.0112	0.19
240	60	0.0042	–.0051	0.085

Thus the change in *t*-butyl bromide concentration over the first 10-second interval is the concentration at the 10 second mark minus the initial concentration at $t = 0$. The average rate over this interval is calculated by dividing the concentration change by the duration of the interval, *Dt*. Rate versus [*t*-BuBr] in a and Rate versus [t-BuBr]^2 in *b*.

The linearity confirms that the reaction is first order in t-butyl bromide. The slope of the plot gives $k = 1.36 \times 10^{-2}\ s^{-1}$. This value is in satisfactory agreement with the value obtained from the ln[*t*-BuBr] versus time plot in Example. Even better agreement can be achieved by plotting average rate over a time interval versus the average [*t*-BuBr] over the same time interval. When the above data are plotted in this manner, the rate constant is found to be $1.33 \times 10^{-2}\ s^{-1}$.

Note that the method of finite differences, like the method based on integrated rate laws, uses the full range of concentration-time data, thus avoiding the major problem with the method of initial rates. For the method to be successful, however, it is important that the time intervals over which average rates are calculated be short relative to the total reaction time.

In the example above, the 10-second time intervals used in the early stages of the reaction, when concentration changes most rapidly, are small compared with the overall reaction time of 240 seconds.

As a useful guideline, time intervals should not exceed 5% of the total reaction time; otherwise inaccurate rate constants will result from the finite difference approach.

Electronic spreadsheets enable the calculations required for the finite difference method to be performed rapidly and conveniently using a computer. Further, the various data plots can be made quickly and professionally using the spreadsheet. With this modern tool, calculation by finite difference methods is routine. We will consider one additional example of the approach.

Example: The decomposition of N_2O occurs according to the following equation:

$$2N_2O\ (g) \rightarrow 2N_2\ (g) + O_2\ (g)$$

In a kinetics run, the concentration of N_2O was found to change with time as indicated in the first 2 columns of Table. The numbers in the remaining columns, calculated using an electronic spreadsheet, represent the average $[N_2O]$ over each time interval, squares of the N_2O concentrations, and the average rates of N_2O concentration change over the successive time intervals. The linearity of the plot and lack of intercept in b confirms that the reaction follows the rate law:

$$-\Delta[N_2O]/dt = k[N_2O]^2$$

The rate constant, obtained from the slope of the plot, is 6.0×10^{-3} L/mole-min.

Pseudo-Order Methods

Many chemical reactions obey a mixed second-order rate law of the type shown:

$$\text{Rate} = k[A][B]$$

The reaction is first order in each of two reactants. An example of a reaction governed by this rate law is given.

$$NH_4^+ + NO_2^- \rightarrow N_2(g) + 2H_2O$$

Suppose that a kinetics experiment is carried out by preparing a solution containing similar initial concentrations of the two reactants, NH_4^+ and NO_2^-, and monitoring the appearance of $N_2(g)$ as a function of time in order to measure the rate. Since the two reactants are present in similar concentrations, both substances undergo significant decreases in concentration over the course of the reaction.

The rate of reaction reflects this simultaneous decrease in concentration of both reactants. Consequently, a plot of Rate versus $[NH_4^+]$ ($NH_4^+ = A$) will be non-linear, even though the reaction is first order in NH_4^+, because the Rate also responds to the change in concentration of NO_2^- (B) significantly during reaction.

Although it is possible to integrate obtain a rather complex equation that may then be used to plot the data, a simpler approach to the kinetics study is often possible. In this approach, a kinetics run is performed on a solution that initially contains a substantial excess of one of the reactants, say B.

If the initial concentration of B is more than 10 times larger than the initial concentration of A, the concentration of B will be nearly the same at the end of the reaction as at the beginning; consequently it may be considered to be constant during reaction. To illustrate this, consider a solution initially containing 0.002 M NH_4^+ and 0.10 M NO_2^-. If reaction is allowed to proceed until all NH_4^+ is consumed, the concentration of NO_2^- will have decreased by only 0.002 M to 0.098 M:

	NH_4^+ +	NO_2^- →	N_2 +	$2H_2O$
initial	0.002	0.10	0	
change	–0.002	–0.002	0.002	
final	0	0.098	0.002	

Thus the concentration of NO_2^- remains essentially constant while NH_4^+ is consumed, and the rate law simplifies as:

$$\Delta[N_2]/dt = k[NH_4^+][NO_2^-] = k'[NH_4^+]$$

Here the constancy of $[NO_2^-]$ has been recognized, and this concentration has been lumped into the rate constant to produce a new constant k'. The rate law now has a simple first-order form, even though the reaction is actually second-order.

This simplified rate law, which may be integrated or subjected to finite difference analysis, is called a pseudo-first order rate law. The prefix, pseudo-, reminds us that the reaction is not truly first-order, but has only been made to appear that way by imposition of a large concentration of NO_2^-.

If one carries out kinetic runs using several (relatively large) concentrations of [B], a pseudo-first order rate constant, k′, will be obtained for each run. Of course, these will have different values depending on [B]. A plot of *k′* versus [B] is then made to obtain the true rate constant, *k*, from the slope.

Example: Pseudo-First Order Kinetics. For the reaction of NH_4^+ with NO_2^-, several kinetics runs were carried out using the initial reactant concentrations in the first two columns of the table below. Plots of Rate versus $[NH_4^+]$ for each run were found to be linear, indicating that the reaction is first-order in NH_4^+. The first-order rate constants obtained from these plots varied with $[NO_2^-]$ as indicated in the third column of the table. What is the order of the reaction in NO_2^-? What is the true rate constant for the reaction?

$[NH_4^+]$initial	$[NO_2^-]$initial	Observed 1st order rate constant, k′
0.00896	0.197	7.06×10^{-5}
0.00904	0.395	1.42×10^{-4}
0.00523	0.593	2.13×10^{-4}

Solution: The order of reaction with respect to NH_4^+ is known to be 1, based on linear Rate vs. $[NH_4^+]$ plots. We thus focus on the relationship between the observed pseudo-first-order rate constants and the concentration of $[NO_2^-]$.

Note that for all three runs, the initial NO_2^- concentration is much larger than the initial $[NH_4^+]$, guaranteeing that $[NO_2^-]$ remains essentially constant during the entire course of reaction. A linear plot of *k′* versus $[NO_2^-]$ raised to a power, *n*, indicates that the reaction is *n*-order in $[NO_2^-]$.

We can determine *n* by trying several plots, first with *n* = 1, then, if necessary, with *n* = 2, 3, and so on, until we obtain linearity.

Plots of k' versus $[NO_2^-]$ and $[NO_2^-]^2$. The top plot is linear, so the reaction is first order in $[NO_2^-]$. The overall second-order rate constant, $k = 3.6 \times 10^{-4}$ L/mole-s, is obtained from the slope.

The Experimental Measurement of Concentration

In order to carry out a kinetics study, it is necessary to have an experimental "handle" on at least one of the reactants or products; that is, the change in concentration of a reactant or product must be accompanied by a change in some measurable property of the system. For reactions involving gases, the total pressure of the reaction mixture changes as reaction proceeds, provided that there is a net change in the amount of gas in the system. Thus, the rate of reaction could be studied by monitoring the total pressure with time. Total pressure decreases as reaction proceeds because only 1 mole of gas forms for each 2 moles consumed.

$$2C_4H_6 \rightarrow C_8H_{12}$$

On the other hand, no change in total pressure accompanies because there is no change in moles of gas; reaction rate can not be studied by monitoring total pressure in this case.

$$H_2(g) + I_2(g) \rightarrow 2HI(g)$$

Rather than measuring system pressure at constant volume, one could measure system volume at constant pressure. In either case, the measured quantity is proportional to the number of moles of gas in the system at a particular time during reaction. Pressure (or volume) measurements are particularly convenient when only a single gaseous reactant or product is involved. In this case, the pressure (or volume) provides a direct measure of the amount of that reactant or product.

An alternative to pressure or volume measurements is to use gas chromatography to monitor the amounts of gaseous reactants and products as a function of time. The gaseous reaction mixture is periodically sampled with a syringe and injected into a gas chromatograph.

In the ideal situation, it may be possible to simultaneously determine the amounts of all reactants and products at each sampling time.

For reactions carried out in solution, a number of methods are used to monitor concentration. The acidity of the solution changes during the course of reaction. Provided the reaction is slow enough, the kineticist could periodically "sample" an aliquot of reaction solution and carry out an acid-base titration using a suitable indicator to determine the $[H_3O^+]$ present at the known sampling time.

$$t\text{-BuCl} + 2H_2O \rightarrow t\text{-BuOH} + H_3O^+ + Cl^-$$

Alternately, since ions are produced but not consumed in the reaction, the kinetics could be monitored by measuring the electrical conductivity of the solution as a function of time. The measured conductivity values can then be related to the total concentration of ions present at that time.

Perhaps the most common method for following the time course of reactions in solution is to monitor the absorption of light by the reacting solution. If one of the reactants (or products) has an absorption band in the *UV*-visible, infrared, or NMR region of the electromagnetic spectrum, the change with time of the amount of light absorbed is a measure of the rate of appearance or disappearance of the absorbing substance. The kinetics of reaction can easily be studied by this method.

$$Cr(H_2O)_6^{3+} + EDTA^{4-} \rightarrow Cr(EDTA)^- + 6H_2O$$

The product, $Cr(EDTA)^-$, has an absorption maximum at a wavelength of 542 nm in the visible region of the spectrum. As reaction proceeds, absorbance at this wavelength increases in direct proportion to the concentration of $Cr(EDTA)^-$, providing a "handle" on the reaction kinetics.

ELEMENTARY PROCESSES AND REACTION MECHANISM

A balanced chemical equation is intended to represent the net transformation of reactants to products that occurs when reaction takes place. It does not, and is not intended to, tell us anything about the details of the reaction pathway. In fact, most reactions occur via a series of relatively simple steps, each involving reaction of only a few (one, two, and very occasionally, three) molecules.

Each of these simple steps is called an elementary process. The equation for an elementary process represents an actual collision event at the molecular level. The sequence of elementary processes by which a reaction is proposed to occur is called the mechanism for the reaction. For example, the reaction is thought to occur by the 2-step mechanism.

Overall:

$$2NO_2\,(g) + F_2\,(g) \rightarrow 2NO_2F\,(g)$$

Mechanism:

$$NO_2 + F_2 \rightarrow NO_2F + F$$

$$NO_2 + F \rightarrow NO_2F$$

with the first step proposed to be slow, the second one fast. Each of the two steps in the mechanism is understood to be an elementary process. We can make several important generalizations about the mechanism.

First, each elementary process involves reaction (collision) of only two species, even though the overall reaction involves 3 moles of reactants. Second, F atoms, produced in the first step and consumed in the second, do not appear in the overall equation for the reaction.

Such a species—first produced, then consumed—is called a reaction intermediate. Third, the elementary processes add to give the overall equation for the reaction. This must always be true.

It is important to realise that a mechanism for a reaction is a hypothesis about what goes on at the molecular level, made by the chemist based on her or his experimental observations. The overall equation for a reaction is NOT a hypothesis; it is an experimental fact, observable in the laboratory.

The mechanism in all of its detail is seldom observable; it must be deduced from experiment, must be consistent with all known experimental facts about the reaction, and is subject to change if in conflict with a new experimental fact.

Types of Elementary Processes

There are three types of elementary process, named for their molecularity—the number of reactant molecules that they involve. A unimolecular process involves only one reactant molecule. Two examples of unimolecular processes are given.

$O_3 \rightarrow O_2 + O$ $\qquad$ $I_2 \rightarrow 2I$

A bimolecular process involves two reactant molecules, which may be the same or different. Examples are given. Bimolecular elementary processes are by far the most common type.

$NO_2 + F \rightarrow NO_2F$

$H_2O_2 + I^- \rightarrow H_2O + IO^-$

A termolecular process requires the simultaneous collision of three reactant molecules.

An example:

$$2NO + O_2 \rightarrow N_2O_4$$

Such three-body collisions are extremely improbable. Consequently, termolecular processes are rare. Elementary processes with molecularity greater than three do not occur, due to the overwhelming improbability of collisions involving four or more molecules.

In contrast to the situation for the overall (net) reaction, the rate law for an elementary process may be written directly from its stoichiometry, because an elementary process is an actual, not a net, process.

Thus for the elementary process,

$$A + B \rightarrow C + D$$

the rate law is $-D[A]/dt = k[A][B]$. Unimolecular processes have first-order rate laws.

The rate law is,

$$-\Delta[O_3]/dt = k[O_3]$$

Bimolecular processes have rate laws that are second-order overall; and termolecular processes have overall third order rate laws.

$-\Delta[NO_2]/dt = k[NO_2][F]$

$-\Delta[O_2]/dt = k[NO]^2[O_2]$

In summary, an elementary process of molecularity n has a rate law of order *n*. The converse is NOT true: a first-order reaction is not necessarily either an elementary process or unimolecular.

The difference between order and molecularity is subtle but important. The concept of molecularity is appropriate only in the context of an elementary process, and indicates the number of reactant molecules involved in (undergoing collision) in the elementary step. The order of a reaction is the number to which the concentration of a reactant must be raised in the experimental rate law for the overall (net) reaction.

This rate law, and the orders in it, do not necessarily correspond to a particular elementary step in the mechanism of the net reaction. We will say more about this shortly.

Rate Determining Step

A particular overall reaction consists of a sequence of elementary steps, which usually occur with different rates. Very often one of the elementary processes occurs much more slowly than do the others and limits the rate at which the overall reaction occurs.

This slow step is then referred to as the rate-determining step (rds). The overall rate is equal to the rate of the rds. The fact that the overall reaction rate is limited by the rate of a particular elementary step is the reason why rate laws frequently do not correspond with the stoichiometry of the overall reaction.

A familiar example of an overall process occuring at a rate limited by a slow step. Theater filled to capacity, connected by a set of wide double doors to a smaller lobby, which is linked to the outside via a revolving door. When the movie is over, people begin to move from the theater to the lobby, and from there to the outside.

We suppose that the double doors between theater and lobby are large enough so that 10 people per second can pass through; but that only 1 person per second can exit the revolving door. We then represent the overall process of leaving the theater in terms of 2 steps:

theater → lobby [10/s]
lobby → outside [1/s]
Overall, theater → outside [Rate = ?]

Initially, people pass from theater to lobby at the rate of 10 per second, and begin to exit the revolving door one by one. However, the lobby quickly fills up, because it cannot be emptied nearly as rapidly as it was filled. The process "piles up" behind the slow step, which is passage through the revolving door. The maximum rate at which people can leave the theater is 1 per second, which is the rate of the slowest step in the process.

Rate Law and Reaction Mechanism

Chemists are very interested in the details of how a reaction occurs; i.e., in the reaction mechanism. The approach to the elucidation of the mechanism of a reaction of known stoichiometry involves at least the following steps:

- Do experiments to determine the experimental rate law for the overall reaction;
- Propose a mechanism for the overall reaction, based on the observed rate law;
- Derive the theoretical rate law implied by the proposed mechanism;
- Compare the rate laws in 1 and 3. If they have the same form, it is safe to conclude that the mechanism is consistent with (not in conflict with) the experimental rate law. If they do not have the same form, go back to 2.

Step 2 makes use of the fact that the observed rate law indicates what atoms are involved as reactants in the rate-determining step of the mechanism. This will be made clear in specific examples to follow. It is the most difficult step of the above procedure, requiring extensive experience and well-developed "chemical intuition." At this level of study, we will not attempt to deduce mechanism from the rate law. Rather, our focus will be this: given the form of the experimental rate law, and one or more proposed mechanisms, decide whether each mechanism is or is not consistent with experiment.

Example: For the reaction $2NO_2(g) + F_2(g) \rightarrow 2NO_2F(g)$, the experimental rate law has the form, Rate = $k_{obs}[NO_2][F_2]$. Two mechanisms have been proposed.

These are given below:

Mechanism 1

$NO_2 + F_2 \Leftrightarrow NO_2F + F$ [k_1, k_{-1} fast]

$NO_2 + F \rightarrow NO_2F$ [k_2 slow (rds)]

Mechanism 2

$NO_2 + F_2 \rightarrow NO_2F + F$ [k_1 slow (rds)]

$NO_2 + F \rightarrow NO_2F$ [k_2 fast]

Can either of these mechanisms be ruled out as being in conflict with experiment?

Solution: We examine mechanism 1 first. To obtain the theoretical rate law (implied rate law) for this mechanism, we proceed as follows:

- Write the rate law for the rds. The rate of this step is the same as the rate of the overall reaction.
  The rds for mechanism 1 is step 2.
  $$\text{Rate} = k_2[NO_2][F]$$
  Although this rate law is valid, it is not expressed solely in terms of concentrations of the reactants in the overall reaction. Since these concentrations are the ones we actually observe, we must somehow rewrite the rate law so as to eliminate the concentration of the intermediate, F, which is not generally observable, and replace it with concentrations of reactants in the net reaction. We can accomplish this making use of the preceding rapid steps, which we assume to occur rapidly enough to come to a pseudo equilibrium. So,
- Use the rapid pre-equilibrium to solve for the concentration of the intermediate in terms of known concentrations. Since the first step of the mechanism is in equilibrium, the forward and reverse rates of this elementary process are the same, and we can write
  $$k_1[NO_2][F_2] = k_{-1}]NO_2F][F]$$

Solving for [F] gives,

$$[F] = k_1/k_{-1} \{[NO_2][F_2]/[NO_2F]\}$$

This expression for the concentration of the intermediate can be substituted into the rate law for the rds:

$$\text{Rate} = k_1k_2/k_{-1} \{[NO_2]^2[F_2]/[NO_2F]\}$$

This is the rate law implied by the mechanism. It is the rate law that the reaction should follow if the proposed mechanism is correct. Clearly, though, it is not in agreement with the experimental rate law, which is first order in NO_2, and shows no dependence on NO_2F. The conclusion is that Mechanism 1 can NOT be correct; it is in conflict with experiment.

We next consider Mechanism 2, using the same general approach.

- rite the rate law for the rds, which in this case is the first step in the mechanism.

$$\text{Rate} = k_1[NO_2][F_2]$$

Further work is unnecessary here, because the rds rate law is already expressed in terms of concentrations of participants in the overall reaction. The implied rate law for Mechanism 2 has the same form as the experimental rate law (first order in NO_2, first order in F_2, second order overall), and is therefore not in conflict with experiment.

Can we conclude, then, that Mechanism 2 is the correct mechanism for reaction? The answer is an emphatic NO. If a mechanism does not agree with experiment, it must be wrong; but just because a mechanism agrees with experiment does not mean that it is correct. A mechanism is a theory—it can never be proved.

To underscore this point, consider the following possible mechanism:

Mechanism 3

$NO_2 + F_2 \rightarrow NO_2F_2$ [k_1 slow]

$NO_2F_2 + NO_2 \rightarrow 2NO_2F$ [k_2 fast]

This also is consistent with experiment, giving the implied rate law,

$$\text{Rate} = k[NO_2][F_2]$$

based on the first step, which is the slow one. You might like to assess the validity of yet a fourth possible mechanism, given below:

Mechanism 4

$1/2\ F_2 \rightarrow F$ [fast]

$F + NO_2 \rightarrow NO_2F$ [slow]

We can now appreciate the significance of a statement made earlier: the experimental rate law tells us what atoms are contained in the reactants of the rate-determining step. The experimental rate law in Example., Rate = $k\ [NO_2][F_2]$, reveals that the reactants in the slow step must contain a total of 1N atom, 2O atoms, and 2F atoms.

Based on this, we can immediately rule out mechanism 1, for which the rds involves only one *F* atom; and mechanism 4, for which the same is true. Both mechanisms 2 and 3 have rds's containing the correct numbers of atoms, so neither is in conflict with experiment.

In general, chemists try to propose mechanisms that agree with the experimental rate law and with any other experimental evidence; and that are chemically reasonable. It is the assessment of chemical reasonability that requires experience.

We consider one additional example of mechanism assessment.

Example: Several mechanisms might be considered for the overall reaction

$$2NO + O_2 \rightarrow 2NO_2$$

with rate law, Rate = $-D[O_2]/dt = k_{obs}[NO]^2[O_2]$. The symbol, k_{obs}, where "obs" stands for "observed", is commonly used for an experimentally measured rate constant.

Mechanism 1:

$NO + NO \Leftrightarrow N_2O_2$ [K_1 fast]

$N_2O_2 + O_2 \rightarrow 2NO_2$ [k_2 slow]

Mechanism 2:

$NO + O_2 \Leftrightarrow OONO$ [K_1' fast]

$NO + OONO \rightarrow 2NO_2$ [k_2 slow]

Mechanism 3:

$NO + O_2 \rightarrow OONO$ [k_1 slow]

$NO + OONO \rightarrow 2NO_2$ [k_2 fast]

Can any of these mechanisms be ruled out?

Solution: Mechanism 3 can be eliminated right away, because the reactants in the rds do not contain the correct numbers of atoms. The experimental rate law demands that there be two N atoms and four O atoms in the rds reactants; however, only one N and three O atoms are involved in the rds of mechanism 3. Mechanisms 1 and 2 show the correct atoms in the rds.

You should confirm that the implied rate laws are:

Mechanism 1: Rate = $k_2K_1[NO]^2[O_2]$

Mechanism 2: Rate = $k_2K_1'[NO]^2[O_2]$

both of which are consistent with experiment. Attempts to draw Lewis structures for the intermediates, N_2O_2 and OONO. The intermediate OONO is a free radical (it has an odd electron), and should therefore react very rapidly with the free radical NO. It is thus unlikely that the second step of Mechanism 2 would be slow. Mechanism 1 is preferable, based on chemical reasonability. There is yet a fourth mechanism, involving a one-step termolecular process. It is currently believed that this is the correct one.

Let's summarize what we have just learned about reaction mechanisms:

- The steps in a mechanism are numbered from 1 to *n*. The subscripts on the elementary process rate constants are given the same numbers, with k_n being the forward rate constant and k_{-n} the reverse rate constant for step *n*. The steps must add to give the overall reaction.
- The mechanism often (but not always) involves at least one slow step. This is the rate-determining step.
- Any fast step preceding the slow step is treated as reversible, and assumed to come to equilibrium.

- To obtain the rate law, we
  - Write the rate law for the elementary rds;
  - Eliminate concentrations of intermediates using rapid pre-equilibrium expressions. This is necessary because the concentrations of intermediates are not usually observable.
- That the reaction proceeds in several steps having different rate constants can cause its rate law to deviate from the overall stoichiometry. However, the rate law tells us which atoms are involved in the reactants of the rate determining step, so is strongly suggestive of mechanism.
- Not all elementary step rate constants need appear in the implied rate law for the mechanism.
- Mechanisms can be disproven, but not proven.

KINETIC MOLECULAR THEORY

The rates of chemical reactions increase with increasing temperature, often dramatically. Coal reacts with oxygen at a negligible rate at room temperature, but burns spontaneously with the liberation of heat and light at 500°C.

A sirloin steak "cooks" (oxidizes) very slowly at room temperature, but requires only 20 minutes at 180°C. Reactions are accelerated by temperature increase because the elementary processes by which they occur take place more rapidly.

We begin this section with two questions:

- Why are some elementary processes fast and others slow at a particular temperature?
- Why are reaction rates (rate constants) so dramatically temperature dependent?

We can arrive at answers to these questions by looking in more detail at the collision process.

Requirements for Collision to Lead to Reaction

As discussed earlier, rate is proportional to the number of molecular collisions occuring per unit volume per unit time. For the decomposition of hydrogen iodide to hydrogen and iodine, assuming that [HI] = 0.001 M and a temperature of 500°C, the rate would,

$$2HI\,(g) \rightarrow H_2\,(g) + I_2\,(g)$$

be 5.8×10^4 M/s if all collisions led to formation of products. The observed rate, however, is only 1.2×10^{-3} M/s, 5×10^7 times slower than it would be if all collisions between HI molecules were effective. Our conclusion must be that only a small fraction of molecular collisions lead to formation of products; i.e., are effective.

In general, for the elementary process:

$$A + B \rightarrow C + D$$

Rate = Number of effective collisions/volume-time, which should be proportional to the total number of collisions per unit volume per unit time. We call this total number of collisions Z. We can relate Z to the concentrations of A and B via equation:

$$Z = Z_o[A][B]$$

where Z_o is the number of collisions per volume per time when *A* and *B* have 1 molar concentrations. The number of effective collisions is less than the total number of collisions for two reasons. First, two colliding molecules must be properly oriented if they are to react.

Reaction requires that bonds within reactants be broken and that new bonds form between atoms in the colliding molecules. A bond cannot form between atoms in separate molecules unless they are near each other during collision. Two of the many possible collision orientations of CO and NO_2, which are reactants.

$$CO + NO_2 \rightarrow CO_2 + NO$$

This bimolecular elementary process involves transfer of an oxygen atom from NO_2 to CO. For this to occur, it is necessary that an oxygen atom of NO_2 approach the carbon atom of CO during collision, so that a bond may form between them. The relative orientation brings the nitrogen atom of NO_2 and the oxygen atom of CO together, so cannot be effective in advancing the process. Similarly, many other possible orientations of these two molecules are unsuitable for transfer of an oxygen atom from N to C, so will not be effective. We symbolize the fraction of collisions having the proper orientation of molecules as p, and recognize that p is a small number.

Second, for reaction to occur, it is necessary that colliding molecules have some minimum total kinetic energy. This is called the activation energy, symbolize E_a. In almost all cases, successful reaction requires that existing bonds be broken and new bonds made. For the bond to break, an input of energy at least equal to the depth of the well is required. This energy is drawn from the total kinetic energy possessed by the colliding species. If the kinetic energy is insufficient to overcome the bonding interactions, the reactant molecules separate without change.

Even if the total kinetic energy is sufficient for reaction, the collision orientation may still be inappropriate for reaction. Both factors—sufficient kinetic energy and correct orientation—are necessary for reaction to occur. Assuming that the minimum energy is available and the orientation correct, bonds will be ruptured, new bonds will form with the release of energy, and new molecules will separate from the collision centre. Between reactants and products, both of which are relatively low in potential energy, lies a state of maximum potential energy called the transition state.

The arrangement of atoms that exists in this state is called the activated complex. We may think of it as a "molecule" in which chemical bonds are in flux, some being broken and others being made.

We can depict this situation graphically in terms of a reaction coordinate diagram, which is a plot of potential energy versus the progress of reaction, which is usually called the "reaction coordinate." A reaction coordinate diagram for the bimolecular elementary process. Potential energy is plotted on the vertical axis, with the reaction coordinate on the horizontal axis. The reactants, A and B are located to the left on the reaction coordinate. As reaction proceeds, potential energy rises to the maximum at the transition state, then falls again as bond formation in the product species is completed and the product molecules separate.

The activation energy, E_a, for conversion of *A* and *B* to C and D (the forward reaction) is the difference in potential energy between the transition state and the reactants. This is labelled in the diagram. In drawing this diagram for the generic process, it has been assumed that products have lower potential energy than reactants; that is, that the process is exothermic. Thus *DE* for the process is negative. The activation energy for the reverse process, also indicated in the diagram, is the difference in *PE* between the transition state and the potential energy of C + D. It is clear from the figure that *DE* for the reaction is the difference between the activation energies for the forward and reverse processes.

This is generally true:

E_a(forward) = *PE*(activated complex) PE(*A* + *B*)

E_a(reverse) = *PE*(activated complex) PE(C + D)

E_a(forward) E_a(reverse) = *DE* for elementary process

Every elementary process can be depicted as a single-humped potential energy curve, with an activated complex at the maximum. Letting the fraction of collisions with sufficient kinetic energy to react be *f*, then

Rate = Number of effective collisions/volume-time = *pfZ*

We are now in a position to answer the first of the two questions posed at the beginning of this section. Some elementary processes are slower than others either because they have very restrictive orientation requirements, corresponding to a very small value for *p*; or they have high potential energy barriers (E_a), corresponding to a small value for *f*. Next we consider the effect of an increase in temperature.

The distribution of kinetic energies in a collection of molecules is governed by the Maxwell-Boltzmann distribution. The M – B distribution curve for temperature T_1.

The value of the activation energy for elementary process has been indicated on the graph. The fraction of molecules with kinetic energy greater than E_a is given by the area under the M – B curve to the right of E_a. As temperature is increased to some new value T_2, the M – B curve shifts to the right and flattens, indicating the increase in both the average molecular kinetic energy and the breadth of the distribution.

The high temperature curve is also shown in the figure. Notice that the area under the curve to the right of E_a is much larger for the high-temperature curve, leading to a dramatic increase in the number of effective molecular collisions. Quantitatively, the fraction of effective collision can be expressed as: f = area under curve right of E_a a $e^{-Ea/RT}$

Thus,

$$\text{Rate} = p\,f\,Z = pe^{-Ea/RT} Z_o[A][B]$$

Equation has the form of a rate law if we identify the terms preceding the concentrations of A and B as the rate constant, k:

$$k = pZ_o e^{-Ea/RT} = Ae^{-Ea/RT}$$

where A contains information about the orientation requirements. Taking the logarithm of both sides, called the Arrhenius equation.

$$\ln k = \ln A - E_a/RT$$

The value of the activation energy for a process is obtained by measuring the rate constant at each of several temperatures, and constructing a plot of lnk versus $1/T$. The plot should be linear, with slope = $-E_a/R$ and intercept = ln A.

We are now able to offer an answer to the second of the two questions posed at the outset: reaction rates are temperature dependent because of the requirement of a minimum kinetic energy for effective collisions.

Example: Aqueous hydrogen peroxide and iodide ion react according to the following equation.

$$H_2O_2(aq) + 3I^-(aq) + 2H^+(aq) \rightarrow I_3^-(aq) + 2H_2O$$

The reaction is first order in peroxide and first order in iodide, second order overall. The reaction was studied at several temperatures, with the following results. What is the activation energy for the reaction?

T, °C	k_{obs}, $M^{-1}s^{-1}$
2	3.03×10^{-3}
17	1.0×10^{-2}
50	7.6×10^{-2}

Solution: The activation energy can be obtained from the slope of a plot of ln k_{obs} versus reciprocal Kelvin temperature. The activation energy obtained from the slope is 53.6 kJ/mole.

Overall (Multistep) Reactions

Thus far we have discussed the temperature dependence of a single elementary process. Similar considerations apply to overall (net) reactions consisting of several steps. We make several general statements about reaction coordinate diagrams for net reactions before looking at a specific example.

- The number of maxima (activated complexes) in a reaction coordinate diagram is equal to the number of steps in the mechanism.
- The slowest (rate-determining) step usually has the highest E_a.
- The rate law for the overall reaction indicates the atoms present in the activated complex for the slowest elementary step.
- Intermediates occur at minima in the reaction coordinate diagram.
- The E_a measured experimentally for the multistep reaction is the difference between the potential energy at the highest point in the diagram and the potential energy of the reactants. (This is analogous to the process of rolling a ball over a succession of hills of varying heights; to make it, the ball must have kinetic energy sufficient to get over the highest hill.)

Example: For the reaction

$$2NO_2 + F_2 \rightarrow 2NO_2F$$

the rate law is $-D[F_2]/dt = k_{obs}[NO_2][F_2]$. Propose a mechanism for the reaction, and draw an appropriate reaction coordinate diagram.

Solution: The rate law tells us that the activated complex of the slow step contains two oxygen atoms, two fluorine atoms, and one nitrogen atom.

This suggests the following mechanism:

$$NO_2 + F_2 \rightarrow NO_2F + F \; [k_1 \text{ slow}]$$

$$F + NO_2 \rightarrow NO_2F \ [k_2 \text{ fast}]$$

Because reaction involves the breaking of the $F - F$ bond and formation of a bond between N and F, the structure is a reasonable arrangement of atoms in the activated complex. Finally, There are two "humps" (maxima), one corresponding to each elementary step in the mechanism; the intermediate, F, is located at the minimum between maxima; the activation energy for the overall reaction is indicated; and the overall reaction is exothermic.

The activation energy for the first step is substantially larger than that for the second step, consistent with the first step being rate-determining. Thus the reaction coordinate diagram reveals by its profile which is the slow step in the overall reaction.

Example: The reaction coordinate diagram for an unspecified reaction. From the details of the diagram, say as much as possible about the net reaction to which it corresponds.

Solution: Because there are three maxima in the diagram, the mechanism must consist of three steps. Step 2 must be the rds, because it has the largest E_a. There is one activated complex for each step, for a total of three. There are two minima, indicating two intermediates. The experimental rate law for the reaction should indicate the composition of the activated complex of the second step. The overall reaction is exothermic, because products are lower on the energy axis than reactants.

Finally, the measured E_a is the energy difference between the highest maximum in the diagram and the reactants. It is shown in the diagram.

The temperature dependence of the observed rate law for an overall reaction consisting of several steps is usually qualitatively similar to that for a single elementary process. However, there are complex mechanisms for which this is not the case.

Indeed, examples are known of reactions for which the observed rate constant is either independent of temperature or decreases with increasing temperature.

KINETICS AND EQUILIBRIUM

The connection between the equilibrium constant for a reaction and the rate constants of the elementary steps by which it occurs is an important one. We begin by stating the relationship. We will then demonstrate, for a specific case, that it is true.

The equilibrium constant for a net reaction is the ratio of the product of the forward rate constants for all steps in the mechanism to the product of the reverse rate constants for all steps in the mechanism:

$$K_{eq} = k_1 k_2 k_{3.} .. / k_{-1}\, k_{-2}\, k_{-3.} ..$$

We now show for a specific net reaction and a specific (proposed) mechanism that this statement is true. Consider again the reaction of NO_2 and F_2 to produce NO_2F, reaction:

$$2NO_2 + F_2 \Leftrightarrow 2NO_2F$$

We have written the reaction with the double arrow because we are now interested in the forward and reverse rates at equilibrium. At equilibrium, the overall forward rate and overall reverse rate must be equal:

$$\text{rate}_f = \text{rate}_r$$

This must be true regardless of the detailed pathway by which the reaction occurs. All available experimental evidence indicates that the mechanism is the two-step process:

$NO_2 + F_2 \Leftrightarrow NO_2F + F\ [k_1, k_{-1}]$

$NO_2 + F \rightarrow NO_2F\ [k_2, k_{-2}]$

When the overall reaction is in equilibrium, so must each elementary step be.

Consequently, we have used double arrows in the elementary steps as well. The equality of the forward and reverse rates of the individual steps is expressed mathematically as follows.

$k_1[NO_2]_e[F_2]_e = k_{-1}[NO_2F]_e[F]_e$

$k_2[NO_2]_e\,[F]_e = k_{-2}[NO_2F]_e$

Eliminating the concentration of the intermediate, F.

$$[NO_2F]_e^{\,2}/[NO_2]_e[F_2]_e = k_1 k_2 / k_{-2}\, k_{-2} = K_{eq}$$

The statement made at the outset is thus seen to be true for this specific case: the equilibrium constant is the ratio of the product of forward rate constants to the product of reverse rate constants.

For the general overall reaction occuring by an *n*-step mechanism,

$$K_{eq} = k_1 k_{2.} .. k_n / k_{-1} k_{-2.} .. k_{-n}$$

Objection is frequently made to the method of derivation that we have just used, because the mechanism for is not known with certainty; it is and will remain a hypothesis, however well based in experiment.

If it is found that the two step mechanism is indeed NOT correct, then the relationship is invalid. This is certainly true. However, even if the mechanism above is incorrect, Reaction must occur by some mechanism.

At equilibrium each step in the mechanism must be balanced in rate, enabling us to eliminate concentrations of any and all intermediates just as we did above, to arrive at a modified version. The validity of the general relationship is independent of the details of mechanism.

Thus the intuitively-expected connection between *k* and *K* does indeed exist. In fact, there are many examples in which the equilibrium constant for a reaction has been obtained from kinetics studies. This process can never be reversed however; it is not possible to obtain *k* by performing equilibrium studies.

Before leaving this matter, we make one more very important point. There is a tendency to believe that reactions with large equilibrium constants are fast, whereas those with small equilibrium constants are slow.

However, neither of these beliefs is true. A large value for K_{eq} means that the product of forward rate constants is much larger than the product of reverse rate constants. It does not follow, however, that the forward rate constants are LARGE. It means only that they are larger than the reverse rate constants. $k_1 k_{2.} .. k_n >> k_{-1} k_{-2.} .. k_{-n}$ does not mean that k_1, k_2..., k_n are large in the absolute sense. A large K_{eq} can result from the ratio of a slow forward rate to an even slower reverse rate.

CATALYSIS

A number of chemical reactions that ordinarily occur slowly can be induced to occur more rapidly by the addition of a suitable substance called a catalyst. We will define a catalyst as a substance that speeds up a reaction without itself being consumed or chemically changed in the overall reaction process. This definition is in practice rather restrictive, because eventually, all real catalysts become deactivated by "poisoning" or via irreversible structural changes; thus they cannot function indefinitely.

For our purposes, here, however, this definition will serve. The reactant molecule that is affected by the catalyst is called the substrate. The process by which a catalyst effects its action is called catalysis. Catalysis is one of the most intensely studied areas in science because it is of tremendous biological and industrial importance. We begin our exploration of catalysts by discussing how they work.

The Mode of Catalyst Function

A catalyst functions either by entering into a slow step of the uncatalyzed reaction mechanism, or by creating an entirely new mechanism for the reaction. In either case, it is thought that the catalyzed pathway has a lower activation energy than the uncatalyzed path. We can illustrate catalytic action generically in terms of the following reaction scheme, where *A*, *B*, *D*, and *F* are reactants and products in the overall reaction, and C is a catalyst for the reaction.

$$A + B \rightarrow D + F \text{ [overall]}$$

Uncatalyzed mechanism:

$A + B \rightarrow AB$ [slow]

$AB \rightarrow D + F$

Catalyzed mechanism:

$A + C \rightarrow CA$ [fast]

$CA + B \rightarrow CAB$ [less slow than first step of uncatalyzed mechanism]

$$CAB \rightarrow D + CF \text{ [fast]}$$

$$CF \rightarrow C + F \text{ [fast]}$$

The effect of the catalyst on the activation energy of the slow step of the reaction is shown schematically. The catalyst affects only E_a; it does not affect the energy of either the reactants or products of the reaction.

The intermediate, *CAB*, in which both reactants are bound to the catalyst, is of particular interest. A possible (and fairly common) structural motif for this intermediate. Notice that, although *A* and *B* are bound to different sites on the catalyst, *C*, they are in proximity and can effectively interact. Binding of *A* and *B* to C causes shifts in electron density that may facilitate bond breaking within *A* and/or *B* and bond formation between *A* and *B* or fragments of them. Thus the catalyst provides an organizing centre for *A* and *B*, facilitating their interaction. In this manner the catalyst can overcome the orientation factor discussed earlier.

As soon as the molecule of catalyst is regenerated in the last step of the 4-step mechanism, it may bind another molecule of *A* and proceed once again through the sequence, converting *A* and *B* to D and *F*. This sequence is repeated a large number of times, so that generally only a small amount of catalyst is required to convert a substantial quantity of reactant to product.

Chemists call the process by which the catalyst cycles through the reaction over and over again a catalytic loop. The unbound form of the catalyst, *C*, is placed at the 12 oclock position of the loop, and the various forms in which the catalyst is found are placed more or less evenly around the remainder of the loop.

Arrows show the direction of reaction around the loop, with reactants brought in from outside the loop, and products ejected cut of the loop. This is a very effective visual presentation of catalyst action.

The number of times that a molecule of catalyst cycles through the loop per unit time is called the turnover number of the catalyst. An effective measure of turnover number is moles product produced per mole catalyst present per time. Let's look at a specific example of the effect of a catalyst on a simple reaction, from which we can draw some general conclusions.

Example: The reaction of ethanol (ethyl alcohol) with bromide ion to produce ethyl bromide and hydroxide ion.

$$C_2H_5OH + Br^- \rightarrow C_2H_5Br + OH^-$$

The reaction is quite slow in neutral or basic solution, but occurs quite rapidly in acidic solution because it is catalyzed by the hydronium ion, H_3O^+. Discuss the mechanisms for the uncatalyzed and catalyzed processes.

Solution: The uncatalyzed reaction is thought to occur in a single elementary step, in which bromide ion attacks the hydroxyl carbon atom of ethanol while the hydroxide group simultaneously departs:

$$Br^- + C_2H_5OH \rightarrow \text{activated complex} \rightarrow C_2H_5Br + OH^-$$

$$\text{Rate} = k_{obs}[C_2H_5OH][Br^-]$$

This mechanism is proposed based on the experimental rate law; which is overall second order. The observed rate law for the catalyzed process is

$$\text{Rate} = k_{obs}[Br^-][C_2H_5OH][H_3O^+]$$

The following three-step mechanism is consistent with this rate law if k_{obs} is identified with k_2K_1:

$C_2H_5OH + H_3O^+ \rightarrow C_2H_5OH_2^+ + H_2O$ [fast, K_1]

$C_2H_5OH_2^+ + Br^- \rightarrow C_2H_5Br + H_2O$ [slow k_2]

$2H_2O \rightarrow H_3O^+ + OH^-$

The reaction coordinate diagrams for the uncatalyzed and catalyzed pathways. Several general statements about catalysis are evident from this example.

- Because the catalyst does not appear in the overall equation for the reaction, it does not affect the equilibrium; it affects only the kinetics. The catalyst speeds up not only the forward reaction, but also the reverse, by the same factor. The position of equilibrium remains the same.
- the activation barrier is lower in the catalyzed mechanism. Even though the catalyzed path involves more steps, the overall reaction occurs more rapidly.

- The concentration of the catalyst appears in the rate law, even though the catalyst does not appear as a reactant or product in the overall equation.
- Although a catalyst and an intermediate share the property of not appearing in the overall equation, they are different types of things. First, the catalyst appears in the rate law; an intermediate cannot. Second, the catalyst appears first as a reactant, then is generated later as a product. An intermediate appears first as a product, then later as a reactant.

Examples of Catalysis

We will now briefly discuss several examples of catalysis that are actually used on a mammoth scale in the chemical industry. Each of these processes, in its own way, has a major impact on the quality of our lives.

The Polymerization of Ethylene

Ethylene is a very simple molecule with formula C_2H_4. It is a gas, obtained as a byproduct during the catalytic cracking of petroleum. Under certain conditions, ethylene molecules can be made to join together end-to-end to form very long chain-like molecules of polyethyiene, so called because it consists of many (poly) ethylenes. The process is represented.

$$2n\, CH_2 = CH_2 \xrightarrow{TiCl_3} -(CH_2CH_2CH_2CH_2)_n-$$

Even at high temperature and pressure of ethylene, this process occurs negligibly slowly. In the presence of a small amount of a modified form of $TiCl_3$ (called tickle-3 in the plastics industry), however, it occurs rapidly at only moderately high temperature and pressure. The process is referred to as Ziegler-Natta catalysis after its two coinventors, who jointly received the Nobel Prize in Chemistry in . Since the discovery of this process in the the entire plastics industry has developed and grown to huge proportions.

To this day, the mechanism of Ziegler-Natta catalysis is not fully understood. Mechanistic studies are difficult for several reasons, one of which is that the process is heterogeneous; that is, the catalyst ($TiCi_3$) and substrate (ethylene) are in different phases.

The catalytic process takes place on the surface of crystals of $TiCl_3$, which rapidly become covered with and blocked from view by the resulting polyethylene. Much effort is ongoing in the US chemical industry to develop more efficient and easily handled Ziegler-Natta catalysts.

The Production of Sulfuric Acid

Year after year, sulfuric acid ranks first on the list of the top ten chemical substances produced in the United States: billions of pounds are produced annually. Sulfuric acid is synthesized by the so-called Contact Process, which involves the four sequential steps below:

$S + O_2 \rightarrow SO_2$ [fast]

$SO_2 + O_2 \rightarrow SO_3$ [slow, because it occurs by a termolecular elementary process catalyzed by V_2O_5]

$SO_3 + H_2SO_4 \rightarrow H_2S_2O_7$ [fast]

$H_2S_2O_7 + H_2O \rightarrow 2H_2SO_4$ [fast]

The reaction of SO_3 with water is very exothermic and causes extensive spattering and production of a fine mist of highly acidic water. For this reason, direct reaction of SO_3 with water in the third step is impractical. Instead, SO_3 is bubbled into pure sulfuric acid, with which it reacts smoothly to give fuming sulfuric acid, $H_2S_2O_7$. This can then be treated with the stoichiometrically correct amount of water to give sulfuric acid. The contact process has been so perfected that sulfuric acid is very inexpensive to produce. Consequently, it is used in any industrial process requiring acid. It finds it major uses in the production of phosphate fertilizers; in paper manufacture; in the petroleum industry; in steel production; and in the production of detergents. The importance of these products in our lives is obvious.

CATALYTIC CONVERTERS

For some years now, catalysts have been placed within the exhaust systems of automobiles to reduce the amount of poisonous or otherwise harmful emissions. These noble metal catalysts (based on platinum and palladium) carry out a dual function. First, they facilitate oxidation of carbon monoxide, resulting from incomplete hydrocarbon combustion, to carbon dioxide:

$$CO(g) + 1/2\ O_2(g) \xrightarrow{Pt} CO_2(g)$$

Second, they catalyze decomposition of nitric oxide, produced during engine operation and oxidized rapidly to toxic NO_2 by atmospheric oxygen, to N_2 and O_2. Unfortunately, catalytic converters also facilitate oxidation of SO_2 to SO_3, which is the precursor of acid rain. Low-sulfur petroleum distillates are therefore essential.

The Haber Process: Vast quantities of ammonia are synthesized each year for use as fertilizer. Currently the most efficient process for ammonia synthesis is the Haber-Bosch Process, developed during the ten-year period preceding in which nitrogen and hydrogen react directly at high temperature and pressure and in the presence of an activated iron catalyst to form ammonia.

$$N_2(g) + 3H_2(g) \xrightarrow{Fe} 2NH_3(g)$$

A catalyst and high temperature are necessary to cause the reaction to go at a reasonable rate. Unfortunately, high temperature makes the exothermic reaction less favored, so very high pressure is used to favour products.

Even with conditions optimized, reaction is incomplete, and unreacted hydrogen gas is recycled for maximum efficiency of ammonia production. Nitrogen fertilizers produced from ammonia are largely responsible for the incredible growth in agricultural production over the decades since the First World War, when the process was first put on line in Germany.

Ironically, the original motivation for development of the process was the requirement of explosives for the war effort. The processes discussed above have at least three features in common. First, catalysis is heterogeneous; the catalyst is in all cases a solid, interacting with the substrate in the gas phase. Second, the mechanisms for these processes are incompletely understood.

Thus catalysts are used successfully on a huge scale, even though we do not understand how they work. Third, in all cases the catalyst involves a transition metal, from the D block of the periodic table.

Transition metals are often versatile catalysts because they are flexible in coordination number (that is, the number of atoms, ions, or molecules to which they may bind in a Lewis acid-base interaction); in stereochemistry (that is, in the shapes that their adducts assume); and in oxidation state (that is, in the charge that they carry). Nature has chosen transition metals to serve as the centerpieces for many of its catalysts, the enzymes, most probably for these same reasons.

Index

D

E

F

G

H

I